THE AUSTRALIAN

# SCUBA DIVERS

## ILLUSTRATED DICTIONARY


ROBERT BERTHOLD



Robert Berthold Photography
15 Wolfe Rd., North Ryde,
NSW 2113 Australia.

Information received from societies, organisations, and other groups has been printed as received and is highlighted in blue.

There are new inventions, organisations, equipment and other information which would enhance future editions of this publication and I invite readers to submit information they would like to see included. If readers find any errors I would like to hear from them as well.


**ISBN 0 646 07526 8**



Published by Robert Berhold Photography
15 Wolfe Rd., North Ryde, NSW 2113 Australia.
Book design and cover photography by
Robert Berthold
Cover design by Anthony van Leuven
Typeset in 10pt Garamond by Character & Caps,
59 Victoria Street, North Sydney.

# Foreword

Australia is favoured by one of the most diverse coastlines of any country in the world. Additionally, because of our comparatively small population per unit length of coastline, we have not suffered the degradation of the marine environment common among the more industrialised nations. Our waters contain some of the world's most attractive areas, such as the Great Barrier Reef, the reefs of the north west of Western Australia, the great diversity of the coast of southern Australia and the magnificent headlands and isolated islands of so many of the coastal areas of eastern Australia.

It is little wonder that scuba diving has attracted so many different sectors of the Australian public to explore the enormous variety of species and habitats in this diverse marine environment where much of the waterways are at a comparatively pleasant diving temperature.

This wide diversity of personal interests almost certainly ensures that no person can dive in any area of Australian waters and be confident of being able to identify the different species that they observe or be knowledgeable on the specific history of the area.

Robert Berthold has been diving professionally and in an amateur capacity for more than 21 years. He is an avid photographer and his collection of photographs is the envy of many scientists and recreational divers. Robert has brought together definitions of words that most scuba divers will have used themselves on at least one occasion, and perhaps never fully communicated to those listening to their adventure stories the meaning of those words. There are, of course, omissions because the complete dictionary for scuba divers cannot yet be written, given that there are still many species still to be identified. This situation adds to the adventure of scuba diving, and also adds to the value of Robert Berthold's illustrated dictionary. I suspect that for many divers it will become effectively a work book where entries can be made for new words that individuals think could be included and when Robert produces his second edition, in perhaps five years, I am sure the contributions of many Australians will result in a considerably expanded version.

The dictionary is recommended to people in all walks of life. Scientists will find it valuable, as will the weekend diver and the dictionary will serve to enhance the readers general knowledge of the marine environment.

As a member of one of Australia's very active scuba diving groups, Robert brings to this dictionary not only his own sense of experience but also that of many colleagues. It is a book which fills a gap in the available literature and one that can be confidently commended for the accuracy of its entries.

The dictionary has added value in that the entries not only describe organisms, events, or organisations but it also provides points for further information on organisations relevant to the interests of scuba divers.

J. T. Baker, OBE, FTS, M.Sc, Ph.D., F.R.A.C.I.
Director
Australian Institute of Marine Science
January 1992.

## Acknowledgements

I owe my success to my wife Lyndall who has helped support me financially as well as by her encouragement and understanding over the past six years.

Special thanks also to Penny Farrant, Carol Mehan and Martin Coleman for proofreading and comments on the text.

For assistance in supplying information for the manuscript, I would like to thank: Australian Institute of Marine Science, Australian Conservation Foundation, Australian Littoral Society, Australian Underwater Federation, Beyond International Group, British Sub-Aqua Club, Coast and Wetlands Society, Confederation Mondiale Des Activities Subaquatiques, Diver's Exchange International, Diving Emergency Service, Earthwatch, Greenpeace Australia, Malacological Society of Australia, Maxwell Optical Industries, National Association of Scuba Diving Schools Australasia Inc. (previously called Federation of Australian Underwater Instructors), National Association of Underwater Instructors, Organisation for the Rescue and Research of Cetaceans, Professional Association of Diving Instructors, Project Stickybeak (Dr D. Walker), Scuba Schools International, South Pacific Underwater Medicine Society, Sydney Maritime Museum, Taronga Zoo Aquarium (John West), The Wilderness Society, Wild life Preservation Society of Australia, Wildlife Preservation Society of Queensland and Universal Dive Techtronics.

Robert Berthold

**abalone.** Edible gastropod mollusc and member of the family Haliotidae. Abalone are also known as mutton fish, ear-shell, or paua and have been used as food in many parts of the world for centuries. There are many species—the two commercial species in southern Australian waters are *Haliotis ruber* the blacklip abalone, and *Haliotis laevigata* the greenlip abalone. Abalone feed at night on algae, those living on exposed coasts and reefs grow faster than those in sheltered conditions. Both species grow to a length of 20 cm and can yield up to 500 gm of white meat from the muscular foot. The greenlip abalone lives on open rock faces and in sea-grass beds and takes three years (average) to reach around 10 cm in size. The blacklip abalone grows at about the same rate but prefers the shelter of rocky ledges and crevices. Predators of abalone include: sharks, stingrays, octopuses and crayfish. The abalone shell is often covered with natural growths including sponges and various species of algae which act as camouflage and can make them difficult to locate. The inside of the abalone shell is lined with mother-of-pearl and this can be used in making jewellery and ornaments. In Western Australia, Roe's abalone *Haliotis roei* is also fished commercially.

**The greenlip abalone *Haliotis laevigata.***

**The blacklip abalone *Haliotis ruber.***

**absolute pressure.** Total pressure exerted, from all sources, including both atmospheric pressure and water pressure. Expressed in units of force per unit area.

**abyssal fauna.** Animals inhabiting ocean waters deeper than 2000 metres.

***Acanthaster planci.*** See **crown-of-thorns starfish**.

**acapnia.** A condition in which there is an excessive loss of carbon dioxide from the bloodstream, via the lungs. Acapnia or hypocapnia, as it is also called, can occur following hyperventilation. See **hyperventilation**.

**Actiniaria.** Order of the class Anthozoa containing the sea-anemones. See **sea-anemone**.

**age: determination of fishes.** The age of fishes can be determined from scales removed from the shoulder area. Examination under a microscope reveals growth rings which are counted to determine the age. Scaleless fishes are aged by examination of their otoliths (earstones) which are composed of layered deposits of calcium carbonate. When cut in half otoliths reveal a series of rings that can be counted in order to determine the age of the fish.

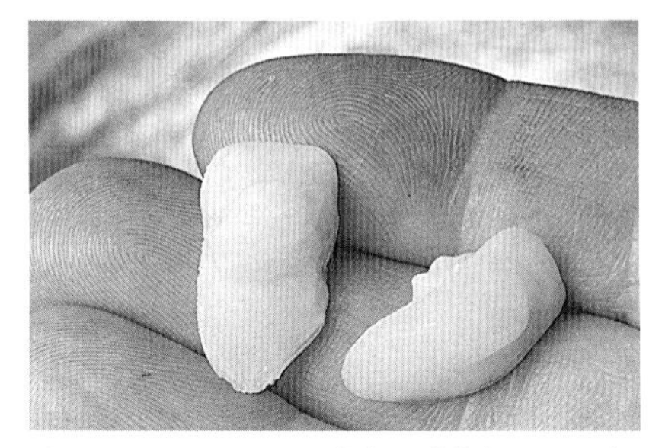

**The earstones or otoliths of fishes can be used to determine their age.**

**AIMS.** See **Australian Institute of Marine Science**.

**air bank.** A number or 'bank' of large air cylinders linked together and charged using a compressor. This arrangement permits more rapid filling of scuba cylinders than with a compressor alone.

**A bank of large compressed air cylinders used for filling scuba cylinders.**

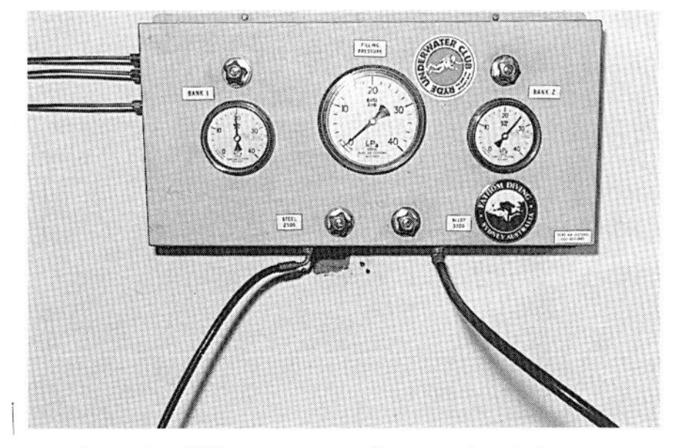

**Scuba air filling control panel with master pressure gauge and high pressure filling lines.**

**air compressor.** See **compressor**.

**air decompression tables.** See **DCIEM Sport Diving Tables**, **RNPL**, **The Wheel Dive Planner** and **US Navy Standard Air Decompression Tables**.

**air embolism.** The obstruction of blood vessels and capillaries by gas bubbles. An air embolism may occur when a scuba diver does not exhale or stops breathing normally when rising to the surface. An air embolism may result in death if cerebral or coronary vessels are blocked by expanding gas bubbles.

**air endurance.** The amount of time a diver can spend underwater using scuba and safely return to the surface with a margin of air in reserve. The duration of a scuba dive, which is directly related to air consumption, is dependent on a number of factors including: air cylinder capacity, depth of dive, emotional state of diver, fitness of diver, individual metabolism, stature of diver, and the amount of exertion undertaken during the dive.

**air lift.** A device which operates underwater with an effect similar to a vacuum cleaner. Air from a surface compressor is pumped to the head of the device where a lever operated by a diver starts and stops the flow of air. The air expands as it rises up a flexible pipe, carrying with it sand, silt and any other small rocks or objects from the sea floor. This device is very useful in clearing large quantities of sand or mud from underwater sites.

**alcyonarian.** Member of the class Anthozoa, subclass Alcyonaria or Octocorallia. See **soft coral**.

**Alcyone.** A ship owned by The Cousteau Society. This radically designed, all aluminium, wind-driven ship, is being used for marine research projects and to test the turbosail system, a wind propulsion cylinder which may revolutionise ocean travel in the future.

**algae.** Simple photosynthetic plants that do not have vascular tissue or produce flowers. The body of an alga is called the thallus and often has root-like formations at the base called rhizoids. There are three main groups of large marine algae or seaweeds; the green algae (division Chlorophyta), the brown algae (division Phaeophyta) and the red algae (division Rhodophyta) which includes the coralline algae. The various types of algae provide food for a host of marine organisms as well as providing: food for humans and livestock, thickening agents for jams, jellies, milk, and cosmetics; fertilisers; agar-agar which is used to support the growth of bacteria in laboratories; alginates which prevent ice-cream from melting too quickly and alginic acid which is used in the manufacture of textiles and adhesives. See **green alga**, **brown alga**, **red alga** and **coralline alga**.

**algal bloom.** A condition where an overgrowth of algae occurs. This can be caused by natural climatic conditions or be brought about by pollution of our oceans and streams with organic wastes such as sewerage and fertilisers. Sewerage in many places is pumped directly into the ocean, while fertilisers and other pollutants soak into the soil and find their way into the watertable or are washed by rain or irrigation run-off into rivers finally ending up in the ocean. Such pollutants are rich in phosphates and nitrates which boost the growth of algae and cyanobacteria (blue-green algae). Algal blooms can be toxic to humans and aquatic creatures alike. See **algae**, **cyanobacteria** and **red tide**.

**ambergris.** A wax-like substance secreted by the intestines and stomach of sperm whales, due to the irritation caused by the hard body parts of squid and cuttlefish. Once highly valued as an aphrodisiac and for the manufacture of perfume; modern synthetic substitutes have been developed. See **whale**.

**ambient.** Immediately or completely surrounding. See **ambient pressure**.

**ambient pressure.** The total pressure exerted by the water and the air above it, upon people or objects placed in it.

**amphora.** A narrow-necked Greek or Roman earthenware vase used mainly for the transport of olives, salted fishes, wines and various oils such as olive oil. Divers' in many parts of the world have recovered amphorae from ancient wrecks.

**ampullae of Lorenzini.** Delicate jelly-filled sensory canals found in the head region of sharks and rays, which help them to locate prey buried under sand. They open to the surrounding water through pore-like openings in the skin and are sensitive to electric fields, salinity and pressure.

**anchor.** A weighted hook with attached rope and/or chain which is dropped to the seabed from a floating vessel to stop it from drifting. See **sea-anchor**.

**anemone.** See **sea-anemone** and **swimming anemone**.

**ankle weight.** Small lead weights used on a diver's ankles to help control buoyancy. Ankle weights are especially useful when wearing a dry suit to help prevent dangerous feet first ascents. See **weight belt**.

**anoxia.** A condition caused by a lack of oxygen, occurring primarily in snorkel divers when they hold their breath too long or when they hyperventilate at the surface before diving. Also called hypoxia. See **hyperventilation**.

**Anthozoa.** The class of coelenterates that includes corals, sea-anemones and zoanthids.

**anti-fouling paint.** Special paint that is periodically applied to boats below the waterline to prevent the growth of marine fouling organisms such as barnacles, bryozoans, algae and wood-boring molluscs. Anti-fouling paint can be toxic to marine life as it generally contains heavy metals and other toxic compounds. Many New South Wales oyster-farmers are facing ruin and blame boat owners who use particularly toxic brands of anti-fouling paint for their losses. The use of the anti-fouling chemical tributyl-tin (TBT) has been banned overseas for many years. The New South Wales Government has banned its use and any other organotin compounds for use on boats under 25 metres in length; the ban took effect from 1 November 1989 and paint manufacturers have been asked to recall stocks held by retailers and wholesalers. A permit is required to use paint containing TBT on vessels over 25 metres in length and all anti-fouling paints have to be registered under the Pesticides and Allied Chemicals Act and the Environmentally Hazardous Chemicals Act. Research carried out by The Cousteau Society in 1985 using its aluminium ship, *Alcyone*, proved conclusively that there are alternatives to using the highly toxic TBT type paints. *Sea Coat* and *XUU 400* anti-fouling paints (both free of TBT additives), proved effective against marine fouling. *Sea Coat* actually proved more effective than the one TBT-containing paint tested, demonstrating that there is no reason for anti-fouling paints containing TBT to remain on the market. *Sea Coat* is an anti-fouling paint made by Donar Chemical Ltd. Canada. XUU 400 is a copper based anti-fouling paint made by International Paint, and is claimed to be non-corrosive when painted onto aluminium.

**antivenom.** A substance which counteracts the toxic effects of a venom.

**anxiety.** A mental state which can increase physical exhaustion and often leads to panic if uncontrolled. Divers who feel anxious should stop whatever they are doing and concentrate on their breathing—slow it down and take longer deeper breaths. This should help control anxiety and prevent it from leading to a panic situation.

**aquaculture.** Cultivation of marine and freshwater plants and animals in artificial ponds and dams on land; and in cages and on artificial surfaces in rivers and estuaries. In Australia, Atlantic salmon, trout and various species of prawns, oysters, mussels and freshwater crayfish are raised on a commercial basis. See **mariculture**.

**Aquadyne.** A brand of commercial diving helmet and related equipment for mixed-gas commercial diving operations.

**Aqualung.** A trade name and a synonym for scuba cylinder and regulator.

**aquanaut.** A person living and working in an underwater research habitat.

**aquaplane.**[1] The act of sliding over the water surface at high speed.

**aquaplane.**[2] A flat plastic or wooden board towed behind a boat and controlled by a diver, allowing large areas of sea floor to be observed, searched or surveyed quickly. Also called a sled. These devices were used extensively in various crown-of-thorns starfish surveys on the Great Barrier Reef.

**Archimedes' principle.** The principle that states: *When a body is wholly or partly immersed in a liquid it suffers a loss in weight equal to the weight of the liquid it displaces.* When a diver floats his/her weight is less than the weight of the water displaced, this is called positive buoyancy. When a diver sinks he/she weighs more than the weight of the water displaced, this is called negative buoyancy. Neutral buoyancy is attained when a diver neither rises nor sinks in the water. See **buoyancy**.

***Artemia salina.*** See **brine shrimp.**

**artificial reef.** A large foreign object on the ocean floor which may provide new fishing and scuba diving sites. The artificial reef may be constructed of old truck tyres wired together, a shipwreck, old car bodies, disused machinery or even old cement blocks and rubble. Any solid object soon becomes encrusted with marine organisms such as sponges, soft corals, algae and bryozoans. Artificial reefs provide hiding places for reef fishes and act as an attractant for pelagic fishes.

**The wreck of the *Coral Queen* off Madang in Papua New Guinea is an excellent example of an artificial reef and it has become a safe refuge to numerous species of fishes.**

**artificial respiration.** A method of restoring respiration in a person who has stopped breathing. This involves an alternating increase and decrease in chest volume while maintaining an open airway in mouth or nasal passages; mouth to mouth resuscitation (EAR) is generally accepted as being the most effective method, but mouth to nose may be more practical for a rescue diver. See **CPR** and **EAR**.

**ascidian.** A member of the subphylum Urochordata, class Ascidiacea, commonly called sea-squirts or tunicates. Ascidian comes from the Greek word *askidion*, meaning a leather wine bag. Simple ascidians are individual animals with a single inhalent and exhalent opening and are usually barrel-shaped (some with stalks) while compound ascidians consist of a number of small individual simple ascidians embedded in a jelly-like matrix which often form irregular encrusting brightly coloured masses. See **compound ascidian** and **tunicate**.

**aseptic bone necrosis.** See **dysbaric osteonecrosis**.

***Asian Diver* magazine.** The first pan-Asian underwater magazine. The first issue was published in June/July 1992. This new dive magazine publishes articles of regional interest to Australian divers. For further information contact: Sports Asia Media Pte Ltd, 19 Tanglin Road, #05-17 Tanglin Shopping Centre, Singapore 1024; phone: 65-7332551, fax: 65-7332291. See **dive magazines**.

**asthma.** A sudden attack of laboured or difficult breathing. The most common form of asthma is allergic asthma characterised by coughing, wheezing, mucoid sputum, laboured breathing and a feeling of constriction in the chest. Under no circumstances should asthmatics take up scuba diving as constriction or narrowing of the airways along with mucoid bronchial secretions while diving could result in a fatal pulmonary barotrauma. See **barotrauma**.

**atmosphere.**[1] The mixture of gases or 'air' surrounding earth.

**atmosphere.**[2] A unit of pressure equivalent to the weight of the atmosphere at sea-level or 101.325 kilopascals. See **pressure conversion table**.

**atmosphere.**[3] An artificial mixture of gases used in underwater habitats.

**atmospheric pressure.** The pressure (or weight) of the air at sea-level, equal to 101.325 kilopascals. See **pressure conversion table**.

**atoll.** A ring-shaped coral reef, with an enclosed lagoon, formed on an existing structure e.g. a submerged volcano. The coral reef grows as the volcanic island is subsiding.

**AUF.** See **Australian Underwater Federation**.

**Australasian Marine Photographic Index (AMPI).** This is a library of colour transparencies showing living marine plants and animals, cross-referenced against identified specimens, housed in various museums and scientific institutes. All transparencies may be hired for illustrations and other commercial uses. The Curator is Neville Coleman, underwater naturalist and author of many books on Australian underwater life. AMPI, PO Box 702, Springwood, Qld 4127; phone: (07) 341 8931, fax: (07) 341 8148.

**Australasian Underwater Photographer of the Year Competition.** The richest competition for underwater photography in Australia. This annual competition is organised by the South Pacific Divers Club, in Sydney. The competition commenced in 1981 as 'The New South Wales Underwater Photographer of the Year Award' and has continued to grow since that time. Today it is recognised as the competition that rewards the best in underwater photography in the Australasian region with the title 'Australasian Underwater Photographer of the Year'. For entry forms write to: Photographic Officer, South Pacific Divers Club, PO Box 823, Bankstown, NSW 2200.

**Australian Conservation Foundation (ACF).** The largest national non-government conservation body in Australia. The following information has been supplied by the Australian Conservation Foundation:

The ACF seeks to protect all of Australia's environment: its land, water, wildlife and atmosphere and works in areas such as: rainforest logging, woodchipping, uranium mining, nuclear disarmament, endangered species, depletion of the ozone layer, the greenhouse effect and climatic change, salinity problems in soil, Antarctica and national parks such as Kakadu, and the Great Barrier Reef.

The ACF emerged in 1963, created by a group of biologists who realised that there was no Australia-wide conservation body that could act as a voice for Australian environmental interests. In 1965 the first Council meeting was held and the first draft constitution was approved. In 1966 the ACF was incorporated as a non-profit association and a year later the first full-time director appointed. The governing body of the ACF is a council of 35 people, elected by the members of their State or Territory. Council meetings take place three times a year at the head office in Melbourne to discuss and develop policies and strategies. The secretariat is based at the head office and directs all administrative, marketing and publishing efforts. There are national campaign offices in all States of Australia, as well as over 20 branches.

The ACF is not aligned politically to any party—it acts only in the interests of the environment, seeking to ensure that wise environmental policies are implemented by all decision makers in government.

The ACF produces a number of publications (*HABITAT* is one) for information and inspiration, and provides an information service for public education.

The ACF is a non-profit organisation financed by membership subscriptions, donations, government grants, bequests, and proceeds from its merchandising and product manufacturing company—ACF Enterprises Pty Ltd. Membership details and other information can be obtained from:

ACF, 340 Gore Street, Fitzroy, Vic. 3065; phone: (03) 416 1455, fax: (03) 416 0767, toll free number phone: (008) 332 510.

**Australian Institute of Marine Science (AIMS).** A tropical marine research institution situated 50 kilometres by road south-east of Townsville at Cape Ferguson, north Queensland. The Institute is a world leader in many fields of marine science, and scientific studies include: biology of the crown-of-thorns starfish, biology of corals, remote sensing, physical oceanography, mangrove and coastal systems, therapeutic substances from the sea, and tropical fisheries ecology. AIMS research is helping to solve the problems associated with the understanding of the complex patterns and dynamic processes involved in our tropical marine systems. Mailing address: Australian Institute of Marine Science, PMB No3 Townsville, MC, Qld 4810; phone: (077) 789 211, fax: (077) 725 852, telex AA47165, cable MARINESCI TOWNSVILLE. See **biomedical marine research**.

**Australian Institute of Professional Photography (AIPP).** Federal Secretary P0 Box 665, Blackwood, SA 5051; phone: (08) 278 1798, fax: (08) 278 3728.

**Australian Littoral Society (ALS).** The following information has been supplied by the Australian Littoral Society:

A national, non-profit organisation concerned with the protection of Australia's rivers, coastal waters and coral reefs. One of its most important roles is to provide information so that people can understand the conservation problems that affect our marine environment. The aims of the Australian Littoral Society are:

- To ensure the protection and wisest use of the Great Barrier Reef.
- To protect our coastal wetlands and promote their importance for fisheries and wildlife.
- To discourage and oppose wanton destruction and/or over-exploitation of aquatic resources, and protect endangered habitats, animals and unique areas.
- To educate the public on the principles and values of effective conservation.
- To provide scientific advice and information to government agencies, private organisations and teaching bodies.
- To investigate the extent and effects of water pollution in Australia.

The Australian Littoral Society's stance on environmental issues is well respected because its statements are always based on the best scientific information available. The Society mobilises the help of concerned scientists to evaluate marine conservation issues and carry out research so that it can present accurate information to politicians and the public.

Since 1968, ALS field teams have been studying threatened mangrove, saltmarsh and sea-grass areas all along the east Australian coastline; surveying large numbers of reefs and islands for their conservation values; designing and helping to build artificial reefs for fishing and skindiving clubs; and helping to solve water pollution problems with projects such as the bringing together of more than 50 experts to prepare a book on the Brisbane River.

One of the Society's most important roles is to provide information to help people better understand the conservation problems affecting aquatic life. Wherever possible ALS members do this by addressing meetings, presenting slide shows and conservation displays, and briefing the news media.

The Australian Littoral Society produces its own bi-monthly Bulletin as well as special publications on important topics such as the Great Barrier Reef, wetlands conservation and management of river systems. Contact: Australian Littoral Society, PO Box 49, Moorooka, Qld 4105; phone: (07) 848 52356.

**Australian Marine Sciences Association (AMSA).** An association of professional marine scientists with the object of promoting liaison between scattered centres and workers in the many disciplines of marine science in all States of Australia through a quarterly bulletin, meetings and conferences, and to promote co-operation between them. Membership is open to scientists, corporate bodies engaged in marine research, or to students of marine science approved by the council of the Association. AMSA aims to improve the public's image of marine scientists and to forward their interests generally. Applications for membership should be sent to: Hon. Treasurer, AMSA, 20/8 Waratah Street, Cronulla, NSW 2230.

**Australian Maritime College.** A college located in Tasmania which offers a three year course for intending fishermen and skippers leading to the Certificate of Technology in Fisheries Operations. The college has designed a range of courses to meet the needs of Australia's shipping and fishing industries—courses which provide 'hands on' experience, using modern and well equipped facilities. Further details from: Australian Maritime College, P0 Box 986, Launceston, Tas. 7250; phone: (003) 260 711, fax: (003) 266 493.

**Australian National Maritime Museum.** This 10-storey waterfront museum was established by the Australian Government to preserve and display Australia's maritime heritage and was officially opened on the 29 November 1991. The Museum is open seven days a week and is located within the Darling Harbour development in Sydney. Australian National Maritime Museum, GPO Box 5131, Sydney, NSW 2001; phone: (02) 552 7777. See **Sydney Maritime Museum**.

**Australian Scientific Divers Association (ASDA).** The Association has a register of over 900 divers working mainly in research and educational fields. For further information contact: Dr Ed Drew, Australian Institute of Marine Science PMB No3, Townsville, Qld 4810. See **scientific diving**.

**Australian sea-lion.** One of the world's rarer species of sea-lion. The Australian sea-lion, *Neophoca cinerea,* is a hair-seal (lacks under-fur) and the males can grow to about 2.5 metres in length and weigh as much as 300 kilograms. There are about 5000 of these mammals living on and around offshore islands from Houtman Abrolhos, Western Australia to Kangaroo Island in South

Australia and their diet consists mainly of fish, octopus, crayfish and squid. Scuba divers in the Marmion Marine Park (only a short drive from Perth) in Western Australia are sometimes lucky enough to swim with these magnificent mammals around Little Island on Marmion reef.

**The Australian sea-lion *Neophoca cinerea* was almost hunted to extinction by sealers in the 1800s.**

**Australian Shark Attack File, (ASAF).** The following information has been supplied by John West, Operations Manager, Taronga Zoo, Sydney:

A file that is held on computer at Taronga Zoo Aquarium. It is associated with the International Shark Attack File which is coordinated by the American Elasmobranch Society.

<u>The Aims and Objectives of the Australian Shark Attack File are:</u>

• To chronicle all known information on shark attacks from Australian waters past, present and future.
• To provide source material for study, to identify the common factors relating to the causes of attacks on humans.
• To provide information held on file for public education and awareness and for publication by the media.
• To publish factual information resulting from analysis of the acquired data.

<u>Criteria for inclusion:</u>

Any human/shark interaction where injury occurs to the human (alive), the equipment worn or held is damaged or where imminent contact was averted by diversionary action by the victim or others (no injury to the human occurred). A detailed questionnaire is available from John West, Taronga Zoo, PO Box 20, Mosman, NSW 2088, Australia.

<u>Research</u>

As part of a worldwide study into shark behaviour, data from the ASAF will help identify the existence, or absence, of common factors relating to the cause of attacks on humans. The research project will be conducted in three stages:

1. To compile as much information as possible on each recorded attack in Australian waters.
2. Assimilated, categorised and transcribed to computer.
3. Analyse acquired data and publish results.

There is a need to learn more about sharks' normal behaviour as well as in circumstances of human contact. This information can only help to accurately predict the behaviour of a shark in a given situation. The ASAF is part of Taronga's marine biology programme and is aimed at understanding and documenting the biology and behaviour of aquatic animals in captivity and in the wild.

<u>The ASAF hopes to dispel common misconceptions such as:</u>

• All sharks are dangerous.
• Sharks only attack when hungry.
• There are lots of deaths from shark attack.
• That sharks lurk off the beaches only to attack, kill and maim humans.

<u>Shark Attack Information</u> (as of 5 December 1991)

Worldwide estimates state that 30 to 100 people a year are attacked. Analysis of over 1000 case histories worldwide, from the International Shark Attack File suggests an average of 30% are fatal.
The Australian Shark Attack File suggests that 39% of local attacks are fatal.

This percentage may come down as more reports of non fatal attacks come into the file from the general public.

The earliest recorded Australian attack was in 1791, the victim was a native female on the north coast of NSW (fatal). The last fatal attack in Sydney Harbour was in 1963, the victim was Martha Hathaway.

Shark Attack Statistics as of 5 December 1991

| State or Territory | Total Attacks | Total Fatal | Last Fatal Attack |
|---|---|---|---|
| New South Wales | 187 | 79 | 1982 Byron Bay. |
| Queensland | 180 | 69 | 1990 off Townsville. |
| Victoria | 19 | 7 | 1977 Mornington Penn. |
| South Australia | 28 | 15 | 1991 Aldinga Beach. |
| Western Australia | 37 | 7 | 1967 Jurien Bay. |
| Northern Territory | 10 | 3 | 1938 Bathurst Island. |
| Tasmania | 16 | 8 | 1982 South Cape Bay. |
| | 477 | 188 | |

Deaths from Shark Attack between 1979 and 1992

Queensland = 6 deaths New South Wales = 1 death Northern Territory = none
Tasmania = 1 death South Australia = 4 deaths Western Australia = none
Victoria = none

The following shark species have been identified in unprovoked attacks in Australia:

White pointer - *Carcharodon carcharias.*
Tiger shark - *Galeocerdo cuvier.*
Whaler sharks - *Carcharhinus* spp.
Wobbegongs - *Orectolobus* spp.

Potentially Dangerous Sharks

Hammerhead sharks - *Sphyrna* spp.
Blue shark - *Prionace glauca.*
Mako sharks - *Isurus* spp.

Any large shark over one metre in length must be considered as potentially dangerous to humans (as must any large animal on land or in the sea).

If you see a Shark

Stay calm and leave the area as quickly as possible. It must be remembered that many stated methods of repelling sharks will, given different conditions and different sized animals, result in altering its initial response and could unintentionally provoke an attack response in the very animal that it was meant to deter. However if an attack is imminent then any action you take may disrupt the attack pattern, such as hitting it, making sudden body movements, blowing bubbles, gouging at its eyes, etc.

If someone is bitten by a Shark

First aid: once the patient is removed from the water;

- Treat the patient immediately. Do **not** rush the patient to hospital.
- Stop the bleeding immediately by applying direct pressure above or on the

wound, a tourniquet may be used if bleeding cannot be controlled by a pressure bandage.
• Reassure the patient.
• Send for an ambulance and medical personnel. Do not move the patient.
• Cover the patient lightly with clothing or a towel.
• Give nothing by mouth.

Prevention of Shark Attacks

Shark attacks remain a genuine, but unlikely, danger when entering the water. This does not mean however that we should disregard the likelihood of an attack by swimming outside the protection of the patrolled beaches (life savers and spotter planes) or protected swimming areas.
People must use common sense as to where they swim and what they do in the water. There is a much higher risk of death from drowning than from an encounter with a shark.

As more knowledge is acquired about sharks' normal behaviour and about the circumstances surrounding attacks, it may be possible to develop an effective repellent. But at this point in time there is little that will act quickly enough to deter a determined attack from a shark. The best prevention is common sense related to where you swim and what activities are undertaken while in the water and knowledge of what may invite or produce an attack:
• Do not swim, dive or surf in areas where dangerous sharks are known to frequent.
• Do not swim alone. Swim, dive or surf with other people.
• Avoid spreading blood or human wastes in the water. Do not swim near a sewer outfall.
• Do not swim in dirty or turbid water.
• Avoid swimming well offshore, near channels, at river mouths or along drop offs to deeper water.
• If schooling fish start to behave erratically or start to congregate in unusually large numbers, leave the water.
• Do not swim with pets and domestic animals.
• Do not swim near people fishing or spearfishing.
• If a shark is sighted in the area leave the water as quickly and calmly as possible.

Anyone wishing to report an encounter with a shark please contact the Australian Shark Attack File Taronga Zoo, PO Box 20, Mosman, NSW 2088; phone: (02) 969 2777, fax: (02) 969 7515.

**Australian Society for Fish Biology.** For information please contact: Hon. Sec. NSW Department of Agriculture, Fisheries Research Institute, PO Box 21, Cronulla, NSW 2230.

**Australian Underwater Federation (AUF).** The following information has been supplied by the Australian Underwater Federation:

The AUF is a national sporting body for underwater activities in Australia. It is supported by individuals through affiliated clubs and their State branches.

It serves scuba divers through providing a competition framework at all levels; representing members' interests to government bodies and the community; and maintaining provision of underwater instruction to World Underwater Federation standards. It is the national representative to the World Underwater Federation—Confederation des Activites Subaquatiques (CMAS). International equivalency certificates (CMAS) are available to qualified divers and instructors through the AUF.

The principal aims and objectives of the Federation are:
• To promote, develop and coordinate all aspects of underwater activities in Australia.
• Organisation of state and national competitions.
• To promote safe practices and equipment for divers.
• To promote the conservation of marine flora and fauna.

Some of the benefits which AUF membership brings to you are:
• Assistance with travel expenses to National Championships.
• An established competition structure for underwater hockey, underwater fishing, fin-swimming, underwater photography and underwater orienteering.
• Eligibility to hold Australian and State records in the aspects of underwater activity.
• Accreditation under the National Coaching Accreditation Scheme for Snorkel and Scuba Coaches and Instructors.
• Representation of your views on government legislation and policy.
• Access to research and advice on diving practices, equipment and emergency support.
• Personal Accident Insurance for divers.
Australian Underwater Federation, PO Box 1006, Civic Square, Canberra, ACT 2608; phone: (06) 247 5554.

**Australian Volunteer Coast Guard.** A voluntary organisation with a vital interest in safety on our waterways. It contributes to the search and rescue operations of official rescue organisations and promotes boating safety by holding training courses in all aspects of safe, small boat operation. Coast guard bases are located in New South Wales, Queensland, South Australia, and Victoria. Details of membership and training courses are available from: The Staff Captain Australian Volunteer Coast Guard, PO Box 30, Oyster Bay, NSW 2225.

**Australian Whale Conservation Society.** A Queensland based cetacean conservation society. All correspondence including membership enquiries, should be addressed to: Australian Whale Conservation Society, PO Box 238, North Quay, Qld 4002.

**back pack or bac pac.** A plastic cradle with quick release shoulder and waist straps which holds a scuba cylinder on the back of a diver. Back packs are usually a built-in feature of modern buoyancy compensators.

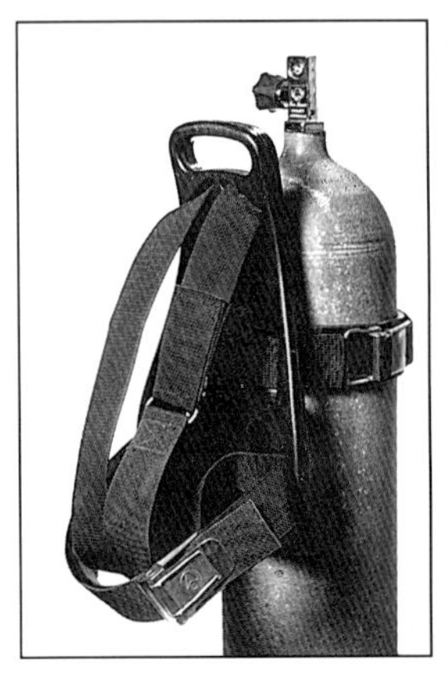

**A back pack is used to hold a scuba cylinder in a comfortable position on a diver's back.**

**Baitweed is collected by fishermen for use as bait.**

**baitweed.** A green alga of the division Chlorophyta. *Enteromorpha intestinalis* or 'baitweed' is a common filamentous green alga on marine rock platforms especially along the New South Wales coastline near Sydney. Baitweed is collected and used as bait for catching luderick *(Girella tricuspidata)*. See **green alga**.

**baleen.** The long flexible plates that hang down from the upper jaw of baleen or whalebone whales suborder Mysticeti. The baleen functions as a sieve, separating the whales' food such as plankton and krill from the seawater. See **whale**.

**bangstick.** An American expression for a powerhead. See **powerhead** and **shark billy**.

**bar.**[1] An offshore reef or sand bank forming a restriction to the free flow of water. In rough weather and at low tide waves usually break over bars making them dangerous to navigate.

**bar.**[2] A unit of pressure equal to 100 kPa. See **pressure conversion table**.

**bar entrance.** A build-up of sand or mud covered by a relatively shallow layer of water, commonly found where rivers flow into the ocean. These areas may be extremely hazardous and a major cause of boating accidents. The inexperienced and the ill-equipped should avoid navigating in these types of entrance. Anyone who must cross a bar should do so only after making enquiries locally as to its condition. Remember the 'wave break pattern' can change according to the tide. Check that the tide is right for both the outward and return trips (ebb tide is the most dangerous, and a crossing should never be attempted at this time). Make sure the boat engine is in good working order.

Watch the weather and listen to forecasts. Have life jackets available and wear them. See **ebb tide**.

**barnacle.** Crustacean and member of the class Cirripedia. There are approximately 900 species of barnacles worldwide, ranging in size from minute forms up to large goose barnacles 80 cm in length. The majority of these crustaceans are attached to rocks by a cement-like secretion and use six pairs of thoracic limbs to comb the water, extracting minute particles of suspended matter for food. Certain species of barnacles live on the skin of whales and sea-snakes, turtles' carapaces, and other marine creatures. Barnacles are notorious fouling organisms on all manner of man-made marine constructions including wharf pylons, oil rig supports and the hulls of ships. See **carapace** and **goose barnacle**.

**The barnacle is a common crustacean of marine rock platforms. Some species seek surfaces where there is high wave action and others prefer less exposed areas in cracks and crevices where there is plenty of shade and moisture.**

**barometer.** An instrument used for measuring atmospheric pressure. There are two kinds of barometer, the aneroid and the mercury. The aneroid is widely used and consists of a flat metal box containing air at a very low pressure. The metal walls of the box are very thin and are held apart by a metal spring. As atmospheric pressure decreases, the spring pushes the sides apart. As atmospheric pressure increases, the sides of the box are squeezed together. These movements are picked up by a series of levers and gears that move a pointer on a calibrated dial on the face of the barometer. Barometric readings are expressed in *millibars* and can be interpreted over a three hour period as follows:

• Barometer reading stable; weather unlikely to change.
• Barometer rising slowly; weather likely to improve.
• Barometer falling quickly (>1 millibar/hour); stormy weather likely.

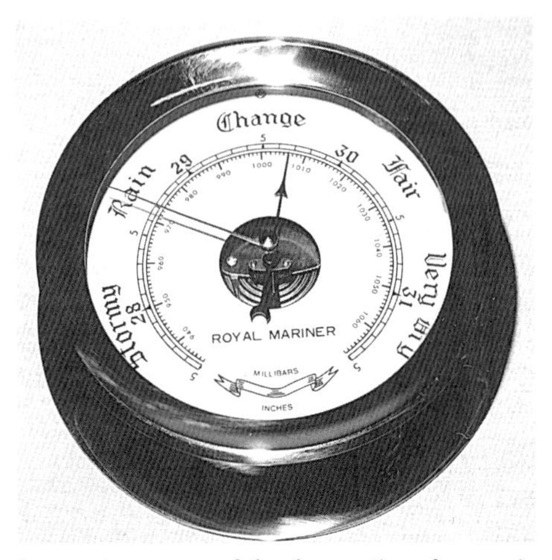

**Barometers are used for forecasting changes in weather.**

**barotrauma.** Damage to the body tissues resulting from large differences between the ambient pressure and the pressure within enclosed air or gas spaces in or around a diver's body. Barotrauma is caused by the failure of the diver to equalise the pressure in the body's air spaces with the pressure exerted by the surrounding water. This can occur when the diver is ascending or descending. The most common type is middle-ear barotrauma and the most dangerous is pulmonary barotrauma of ascent. See **middle-ear** and **pulmonary barotrauma**.

**barracouta.** Fishes belonging to the family Gempylidae, genus *Leionura.* These are large pelagic fishes of the colder southern oceans. The barracouta has two dorsal fins, joined at their bases, followed by five or six detached finlets towards the tail. In contrast the barracuda (family Sphyraenidae) of the tropical north has two separate dorsal fins but no finlets near the tail. See **barracuda**.

**Barracudas of the genus *Sphyraena* are found worldwide in warm tropical seas.**

**barracuda.** Fishes belonging to the family Sphyraenidae, genus *Sphyraena.* There are about 20 species of barracudas living in tropical and temperate seas worldwide. They are large predaceous pelagic fishes, growing up to 2.4 metres in length. Once reputed to attack divers, these fish look more fearsome (with their protruding jaws packed with long sharp teeth) than their reputation deserves. No recorded attacks on humans have occurred in Australian waters but attacks have occurred in the Atlantic. Barracuda will sometimes follow or circle divers, and they are supposedly attracted by shiny objects such as chrome watches. See **barracouta**.

**barrier reef.** Coral reefs separated from the coast by a lagoon. See **Great Barrier Reef**.

**bathyscaphe.** A free diving vessel, consisting of a spherical manned chamber topped by a buoyancy tank filled with an incompressible lighter-than-water fluid and a series of ballast compartments filled with iron or lead shot. See **deepest manned ocean descent**.

**bathysphere.** A spherical steel diving apparatus manned by two persons and their instruments and capable of attaining great depths; used in oceanography for research purposes. The bathysphere having no engine and no means of navigation was tethered by cable to a ship on the surface; through this cable flowed the supply of oxygen which kept the occupants alive. In 1934 off Bermuda, William Beebe and Otis Barton descended in the vessel *bathysphere* to a depth of 925 metres.

**BC, BCD and BCJ.** See **buoyancy compensator**.

**BCD low pressure inflator.** A mechanism used on buoyancy control devices that allows both oral inflation and mechanical inflation using air supplied by a low pressure hose from the first stage of the diver's regulator.

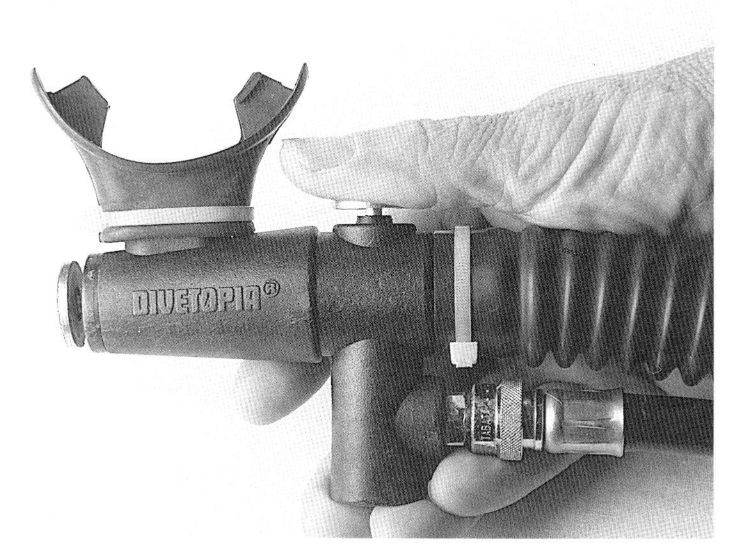


**BCD low pressure inflator.**

**bearing.** The direction of an object from a boat, or another object. Compass bearings express direction relative to magnetic north.

**Beaufort scale.** A scale which groups wind speeds up to 64 knots into forces 1 to 12. Each group of wind speeds is recognised by its effect on the appearance of the sea's surface.

**beche-de-mer.** See **sea-cucumber**.

**bends.** See decompression sickness.

**benthic.** Living in or on the bottom of oceans, rivers and lakes.

**benthos.** Aquatic plants and animals which live in or on the bottom of oceans, rivers and lakes.

**berried.** Carrying eggs on the outer surface of the body, e.g. female crabs.

**bezel.** A rotating collar or ring found on most diving watches, used to tell elapsed time on a dive.

**A diver's revolving watch bezel.**

**bilge pump.** A manually operated, power assisted or automatic pump designed to draw water from within the bilges or lowest levels of a ship. Generally the pump is fitted with a strainer to prevent blockages, and may operate automatically according to the amount of water in the bilge.

**biology.** The study of all living things.

**bioluminescence.** Light emission by living organisms. The light or bioluminescence is produced within the cells of the organism: chemical energy is transformed into light energy during a chemical reaction in which electrons are excited to a higher energy state and emit light as they return to the normal state. Many organisms are capable of producing luminescence including species of: dinoflagellates, comb jellies, corals, worms, shrimps, and fishes.

**biomedical marine research.** Marine research for biomedical application, conducted by government bodies, universities and private research organisations. In Australia the CSIRO Division of Fisheries is researching the role of marine oils in preventing occlusive vascular disease in humans. The Sir James Fisher Centre (part of the University of Townsville, North Queensland) is working on bacterial contaminants in coral injuries. The School of Biochemistry, University of New South Wales is researching polypeptide cardiac stimulants from sea-anemones, while the University of Queensland is working on the chemistry and pharmacology of ciguatera-like toxins from marine fishes. The Australian Institute of Marine Science (AIMS), north Queensland, may give us a more efficient sunscreen lotion as a result of research into the ultraviolet blocking substances found in Great Barrier Reef corals. The National Cancer Institute

(USA) is examining marine organisms as possible sources for new pharmaceutical treatments for cancer and the AIDS virus. See **drugs from the sea**.

**biosphere.** The living world; all the life on earth. This includes all the regions of the earth where life is found.

**bivalve.** A mollusc having a shell in two parts hinged together, such as the clam, mussel, scallop, oyster etc. See **pipi**.

**black coral.** Anthozoan and member of the order Antipatharia. Black corals are colonial animals that form bushy branches or single whip-like strands which may reach five metres in height. The dense branches are black and covered with non-retractable polyps, commonly white, orange or yellow in colour. Black corals are collected commercially, cut and polished, for use in jewellery.

**blind shark.** A small harmless shark *Brachaelurus waddi,* member of the order Orectolobiformes, family Brachaeluridae. The first recorded catch of a blind shark was by a member of the First Fleet in Sydney Harbour. The common name is derived from the shark's habit of completely closing its eyes when taken from the water, leaving only two small slits visible on the head. A harmless and sluggish species, the blind shark grows to around 120 cm in length and lives on the bottom under rocky ledges in shallow water from southern Queensland to Jervis Bay in New South Wales. It feeds mainly at night on crabs, and other crustaceans.

**Juvenile blind sharks *Brachaelurus waddi* have a series of brown bars on their backs and sides, which fade as they reach maturity.**

**bloom.** See **algal bloom** and **red tide**.

**blowout plug.** See **burst disc**.

**bluebottle.** See **Portuguese man-of-war**.

**blue-green algae.** See **cyanobacteria**.

**blue groper.** Fish belonging to the family Labridae (wrasses). Blue gropers are wrasses and are not related to any of the true Australian gropers, (family Serranidae). The common species scuba divers are likely to see are: the eastern blue groper *Achoerodus viridis,* which inhabits rocky coastlines off southern Queensland, New South Wales and Victoria, where they grow to about one metre in length; and the western blue groper *Achoerodus gouldi,* which is found in waters off Victoria, South Australia and Western Australia, where they grow to about 1.6 metres in length. In areas where they are not hunted blue gropers will readily take a bait from the hand of a scuba diver. See **Queensland groper**.

**The male eastern blue groper (wrasse) *Achoerodus viridis* grows to more than a metre in length and may weigh as much as 49 kg.**

**blue-ringed octopus.** A mollusc and member of the class Cephalopoda, family Octopodidae. Two species of blue-ringed octopus are well known: the southern blue-ringed octopus *Hapalochlaena maculosa* found in southern Australian waters and growing to about 12 cm across the arms; and the northern blue-ringed octopus *Hapalochlaena lunulata* the larger of the two species growing to 20 cm across the arms and found in northern Australian waters. These small cephalopods live in rock pools and areas of sea-grass and are commonly found hiding in discarded sea-shells. In polluted waters such as Sydney Harbour they often live in old beer cans and bottles. Their diet consists of small crustaceans such as shrimps and crabs. Do not handle or provoke a blue-ringed octopus as its bite is venomous and the fast acting neurotoxin can cause death from respiratory failure within minutes. If bitten, the best first aid procedure is to apply a pressure bandage and immobilise the limb. Reassure the patient, even if totally paralysed, as he/she will still be fully conscious and able to hear. If the patient's breathing stops, commence mouth to mouth resuscitation (EAR) and continue until the patient is hospitalised and placed on a respirator. After treatment the paralysis subsides and the respiratory muscles begin to function properly again in about 8-24 hours.

**The southern blue-ringed octopus *Hapalochlaena maculosa* is regarded as potentially one of the most dangerous marine creatures in Australian waters.**

**blue swimmer crab.** A crustacean and member of the order Decapoda. *Portunus pelagicus* commonly known as the blue swimmer crab, manna crab or sand crab, grows to one kilogram in weight, measures up to 20cm across the carapace (shell) and can be found in most bays and estuaries around Australia, being eagerly sought by amateur and professional fishermen alike. See **carapace**.

**The blue swimmer crab *Portunus pelagicus* can be found in most bays and estuaries around the Australian coastline.**

**boat diving.** Scuba diving from a boat rather than from the shore. In Australia most scuba diving is done from boats operated by professional dive shops. Here are a few hints for safe boat diving: use a small buoy on a length of line (also called a 'mermaid catcher') to check for currents and to act as a safety line, erect the International Code 'A' flag (diver below flag), swim down the anchor line and secure the anchor, always swim into the current and always leave a competent person in the boat to help any diver in distress on the surface. You should also carry a first aid kit and a supply of oxygen as well as the normal items of boating safety equipment as required by the maritime regulations in each State. See **boating safety equipment**.

**boating safety course.** A course that will qualify you for the Maritime Services Board Boat Driver's Licence and the Australian Yachting Federations Boat Handling Certificate. The course consists of 27 hours of lectures on the following subjects: seamanship, preventing collisions, navigation, emergencies, meteorology, safe boating, boat handling—power and sail, buoys and markers, accidents, and motor maintenance. Contact the Course Information Officer at your nearest TAFE college or MSB Boating Safety Access Line; phone: (02) 364 2888 or (008) 422 718 (toll-free outside Sydney area).

**boating safety equipment.** The minimum safety equipment that you should carry when in open water on a small boat. The equipment includes a marine radio transmitter, life jackets for everyone on board, flares, auxiliary motor, bucket, compass, oars, fire extinguisher, fresh water and 'V' distress signal sheet. If possible also include a life boat or raft and an Emergency Position Indicating Radio Beacon (EPIRB). See **EPIRB** and **distress flare**.

**boating terms.**

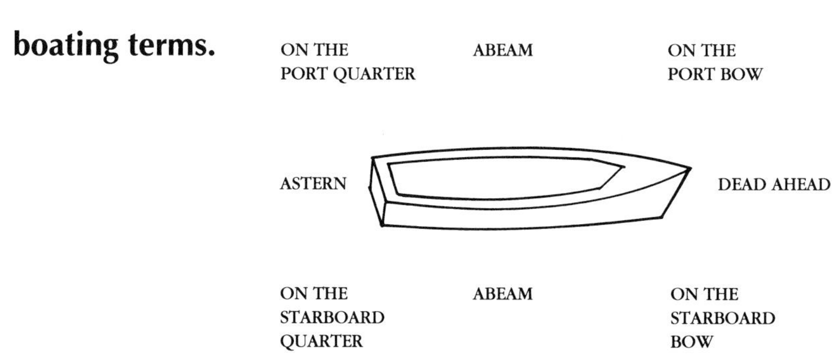


**boating weather forecast.** A recorded information service giving details of local weather conditions which might affect boat operators. Look in your telephone directory under 'recorded information services'. Sydney boating enthusiasts should phone: (02) 11541.

**boat ramp.** A gently sloping structure on the water's edge, most commonly constructed from concrete—used by boating enthusiasts for launching and retreiving trailer boats.

**bollard.** A short post on a boat or wharf, usually made of wood or metal and used for securing mooring lines from the boat to the wharf or vice versa.

**bommie.** A small isolated patch of coral reef which is higher than the surrounding reef platform or bottom and which may be partially exposed at low tide.

**bone necrosis.** See **dysbaric osteonecrosis**.

**bone rot.** See **dysbaric osteonecrosis**.

**bookshops for divers.** Specialist bookshops for scuba divers include:

• HELIX: The largest underwater photo outfitters in the world. Stocks a huge selection of underwater books and photographic equipment, write for a free catalogue to: HELIX, 310 South Racine Avenue, Chicago, Illinois 60607, USA; phone: (312) 421 6000, fax: (312) 421 1586.

• Ocean Enterprises: A free catalogue of over 200 specialist diver's books is available from Ocean Enterprises, 303-305 Commercial Road, Yarram, Vic. 3971; phone: (051) 825 108, fax: (051) 825 823.

• Underwater Geographic Bookshelf: A mail order service operated by Neville Coleman, renowned naturalist, author and underwater photographer. Contact: Underwater Geographic Book Shelf, PO Box 702, Springwood, Qld 4127; phone: (07) 341 8931, fax: (07) 341 8148.

**boot.** Plastic or rubber cup-shaped receptacle fitted to the base of a scuba cylinder to protect it from abrasions and enable round-bottomed steel cylinders to be stored in a vertical position.

A boot made of plastic or rubber is used to protect the bottoms of scuba cylinders from damage.

**booties.** Shoe-like coverings for the feet and ankles made from neoprene rubber, which help to insulate the wearer from cold water and protect them from chaffing when wearing swim fins. There are soft sole varieties for wearing with enclosed heeled fins and hard sole types for open heeled fins.

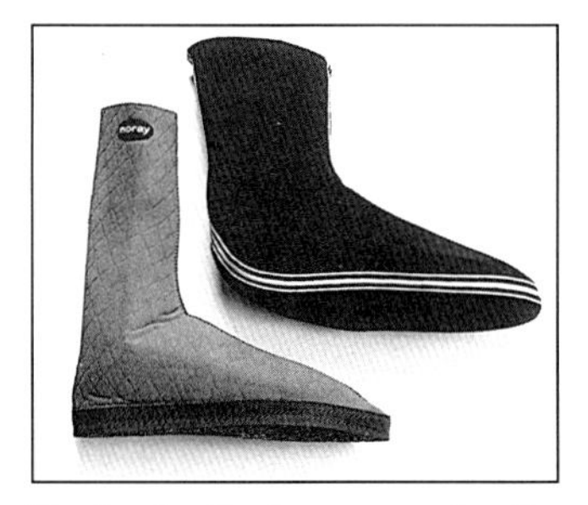

Hard and soft sole neoprene booties (soft sole at top).

**bore.** A high wave of water moving upstream, caused by an incoming tide opposing a river's normal current flow; usually occurring at the mouth of a river.

**bottle.** Term used for a compressed air cylinder of any capacity.

**bottom time.** The total lapsed time between leaving the surface and beginning the ascent. It is important to remember when reading your dive tables that bottom time is not just the time spent at the maximum depth.

**bottom timer.** A diving instrument that automatically starts and stops at a specified depth, usually 1–2 metres. Its main advantage compared to a diver's watch, is that it operates on a completely automatic basis giving lapsed bottom time in minutes. Using a diver's watch the diver has to remember to set his/her watch bezel manually in order to record bottom time. The latest 'combo' dive gauges include bottom timers.

**Boutan, Louis.** A French scientist and pioneering underwater photographer who in 1893 successfully took underwater photographs while working at the marine laboratory at Banyuls-sur-Mer. Boutan also published a book in 1900 titled *La Photographie Sousmarine.* This was probably the first book devoted to underwater photography. See **underwater photography books**.

**box jellyfish.** Coelenterate and member of the order Cubomedusae. There are two families; the Carybdeidae which have four tentacles and the Chirodropidae which have four clusters of tentacles. Box jellyfish were formerly called 'sea wasps'. Several species can cause stings but *Chironex fleckeri,* which is commonly found in the muddy coastal waters and tidal streams of northern Australia and the Indo-West Pacific, is the most dangerous jellyfish in the world. This jellyfish has caused approximately 70 deaths over the years, although an antivenom available since 1970 has saved many lives. Clothing such as nylon pantihose or stinger suits made from lycra, or in fact any clothing which adequately covers exposed skin will prevent stings. Weak acetic acid (vinegar) is the safest material to carry in a first aid kit to treat all jellyfish stings. Vinegar should be poured liberally onto the adhering tentacles to inactivate them and the tentacles can then be removed gently. Mouth to mouth resuscitation (EAR) should be commenced if the patient's breathing stops and cardiopulmonary resuscitation (CPR) started if the pulse ceases. See **lycra suit**.

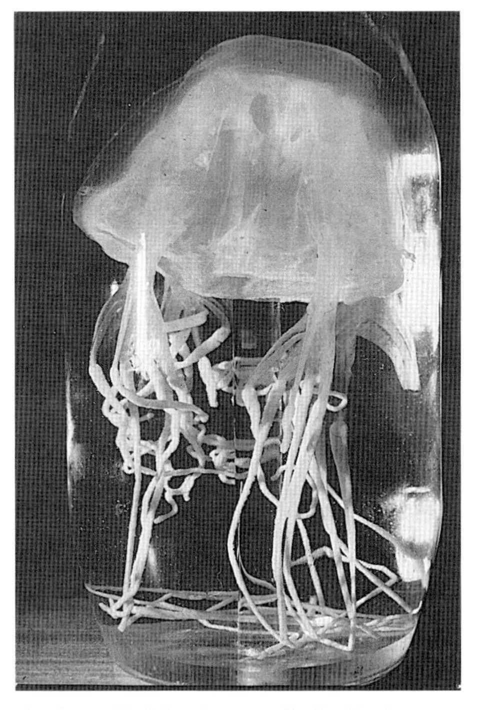

**The box jellyfish *Chironex fleckeri* is the most dangerous jellyfish in the world and has been the cause of a number of bather's deaths.**

**Boyle's law.** A law formulated by Robert Boyle (1627–97), which states that: *If the temperature remains constant, the volume of a given mass of gas is inversely proportional to the absolute pressure.* A practical example of Boyle's law occurs when a snorkel diver holds his/her breath and descends. The volume of air in the lungs of the diver will be reduced as the ambient water pressure increases. During ascent the pressure decreases and the air in the diver's lungs expands. Upon reaching the surface the air in the diver's lungs will be the same volume and pressure it was at the commencement of the dive.

**bradycardia.** Slowing of the heartbeat. In humans and other mammals this occurs involuntarily when the face is immersed in water.

**breakwater.** A structure, usually of rocks or concrete which extends from the shoreline and breaks the force of waves. Breakwaters are used to enclose and protect marinas and artificial harbours from the forces of the sea.

**breathhold deep diving record.** The Italian Angela Bandini dived with the aid of a weighted sled to a depth of 107 metres (351 feet) off Elba, Italy, in October 1989.

**breathhold diving.** Diving without the use of scuba, with or without the benefit of mask, snorkel and fins.

**brine shrimp.** Tiny free swimming crustacean and member of the class Branchiopoda. Brine shrimp, *Artemia salina,* live in salt lakes in many parts of the world but not in the ocean. They are coloured bright red and grow to 15 mm in length. Their eggs can survive long periods out of the water (2–3 years) because their thick egg cases prevent dehydration and as a result of this, eggs are sold in many countries to aquarists who hatch them and use the tiny shrimp as food for a variety of aquarium fishes and invertebrate species.

**bristle worm.** Marine annelid worm and member of the class Polychaeta. Each body segment of the bristle worm contains paired bunches of long white bristles which are very sharp and brittle. When touched, the bristles easily penetrate human skin even leather or rubber gloves and break off. They are almost impossible to remove. Adhesive tape sometimes helps in removing the most obvious bristles and application of vinegar helps to dissolve them. These worms are commonly found under rocks in the intertidal zone.

**British Sub-Aqua Club (BSAC).** A diving club formed in London in 1953 by Oscar Gugen and Peter Small. The following information has been supplied by the British Sub-Aqua Club:

The BSAC is a non-profit organisation that aims to promote safe diving as a sport for those interested in the underwater world. It operates from headquarters in London and is governed by a committee of elected representatives who are responsible for setting the standards and policies to be implemented by all branches of the BSAC worldwide. These standards include a

**Breathhold diving, also called snorkelling, can be a fun way of exploring reefs in clear shallow water.**

diver training programme from novice through to advanced diver and instructor qualifications, which are supported with a comprehensive range of syllabi and teaching aids aimed at maintaining a minimum standard of training and certification. This training is recognised by the CMAS around the world as meeting its necessary requirements. The BSAC qualifications are therefore recognised worldwide.

Approximately 12 BSAC branches operated around Australia in 1990, each with a committee elected from within its own membership. In addition to training, each branch operates a regular dive programme run by experienced dive marshals who supervise the dives within the safety standards set down by the BSAC. These branches are a source of common interest both from a sporting and social viewpoint, for divers and their families and provide members with safe and economical diving. Each branch organises activities to suit their members' requirements, but all are controlled and maintained on a voluntary basis. Branches rely on all members to chip in and support in any way possible. This may take the form of fund raising, gaining instruction qualifications, boat handling, equipment maintenance and the list goes on. There are numerous BSAC branches worldwide, so members are able to visit and dive with these branches, and still obtain the benefits of cheaper, safer diving.

Because of the voluntary status of BSAC Instructors, diver training courses tend to take longer than commercial counterparts. This is often an advantage to the student as it gives ample time to gain experience and confidence while obtaining qualifications. The BSAC training program is widely recognised and acclaimed for its thoroughness and safety. It commences with the use of fins, mask and snorkel and progresses to aqualung training in the pool. When ready, the novice progresses to open water where he/she carries out a number of qualifying dives. As an example, a typical novice, sports diver course operated within a branch would take approximately three months to complete.

So why join a club? Besides the advantage of receiving sound instruction, diving is not a sport that can be carried out safely alone or in very small groups, unless the scope is very limited. Dive partners are required, safety cover, boat crew, shore control etc. Two or three individuals cannot carry out adventurous diving without adequate support. A club also provides such facilities as equipment, advanced training, pool sessions to maintain fitness, and by no means least, companionship and an enjoyable social atmosphere.

The member's subscription provides him or her with a copy of the BSAC diving manual, monthly issues of the BSAC magazine, comprehensive third party and public liability insurance cover (currently approximately $3 million to all members for all diving activities including guests to the club), and many other advantages. The BSAC in Australia is like any sporting club—you can get a lot out of it if you want to, but only if you are prepared to put something into it. Contact: British Sub-Aqua Club, PO Box 318, Noble Park, Vic. 3174.

**brittle-star.** Echinoderm and member of the class Ophiuroidea. The flattened body consists of a central disc with five or more long slender flexible arms.

Brittle-stars can move surprisingly quickly and are mainly detrital feeders. It is difficult to pick up one of these starfish without it losing a number of the arms—hence the name brittle-star. See **echinoderm**.

**Brittle-stars should not be handled as their long slender arms are easily broken.**

**broach.** When a ship veers to windward it is said to broach and is positioned broadside to the wind.

**brown alga.** Member of the division Phaeophyta. Brown algae vary from yellow and brown to almost black in colour. This group contains the largest of all marine plants, *Macrocystis pyrifera,* also known as giant kelp, a seaweed which is confined to south eastern Tasmania where it may grow to 20 metres in length. Along the Pacific coast of California this same species may reach an incredible 100 metres in length. Brown algae are conspicuous in the lower eulittoral zone, and are common inhabitants of intertidal rock pools. In southern Australia, intertidal pools are often fringed by *Ecklonia radiata, Hormosira* *banksii,* and species of *Cystophora* and *Sargassum.* Brown algae are abundant in the subarctic and temperate oceans but few large species are found in tropical waters. See **kelp**.

The brown alga *Hormosira banksii* is commonly called 'bubble weed' or 'Neptune's necklace'. It was first noted in Australian waters by the botanist Sir Joseph Banks, a member of the First Fleet.

The brown alga *Macrocystis pyrifera* also called giant kelp is confined to south eastern Tasmania where it grows to about 20 metres in length.

**bryozoan.** Member of the phylum Bryozoa. There are about 4000 species of bryozoans. Individual animals are small, ranging from one millimetre to several millimetres. However these tiny animals form sessile colonies which can grow to more than a metre in height or diameter. Colonies can take the form of flat crusts, ears, tufts, lobes or lacy coral-like growths, and can be found living on rocks, algae, sponges or for that matter any other submerged object, living or dead. They

The bryozoan *Celleporaria* sp. grows mainly in the form of erect blades. This specimen was photographed at Portarlington, Victoria.

often cause major fouling problems on the hulls of ships and on the support legs of offshore oil rigs.

**This bluish-green bryozoan *Bugula dentata* is growing on a sponge, behind it is the orange coloured bryozoan *Orthoscuticella* sp.**

**BSAC.** See **British Sub-Aqua Club**.

**buddy breathing.** A technique by which two divers share a single air supply when one diver's air supply is depleted or faulty. This can be accomplished by sharing the air from a single regulator mouthpiece or by the more popular method of using a second regulator mouthpiece attached to the first stage, commonly called an 'octopus' regulator. See **octopus regulator**.

**bull-kelp.** See **brown alga** and **kelp**.

**buoyancy.** The upward force exerted on a diver by the surrounding water.
• Positive buoyancy: the diver rises to the surface.
• Negative buoyancy: the diver sinks to the bottom.
• Neutral buoyancy: the diver remains at the chosen depth without effort.

**buoyancy compensator.** A device used to aid in ascents, maintain depth profiles and for positive buoyancy at the surface. Known by various acronyms such as BCD, and BCJ. The three main variations are: the horse collar which fits over the head, e.g. the Fenzy and Avon types; the jacket type (incorporates a back pack and inflates in the shoulder area) and the vest type (incorporates a back pack and has padded straps over the shoulders and most of the buoyancy under the arm area). Most models feature at least two methods of inflation, including: oral inflation, scuba feed, carbon dioxide cartridge or inflation from a small independent air cylinder. See **Fenzy**.

A buoyancy control device or BCD is an essential item of scuba equipment.

**buoyancy control device (BCD).** See **buoyancy compensator**.

**buoyancy control jacket (BCJ).** See **buoyancy compensator**.

**burrowing clam.** See **clam**.

**burst disc.** A small copper safety disc fitted to the on off valves of scuba cylinders. The burst disc is designed to burst if the cylinder is over filled or over heated allowing the air to escape in a safe way and protecting the cylinder from exploding.

**butterfly cod.** See **zebra lionfish**.

**by-the-wind sailor.** Floating oceanic coelenterate and member of the class Hydroza, order Chondrophora. *Velella velella* or by-the-wind sailor is blue in colour and consists of an oval float about 1-5 cm in length, bearing a sail which projects across the float at an angle. It has a single central feeding polyp surrounded by reproductive polyps and feeds mainly on planktonic animals and small shrimps. Predators of by-the-wind sailors include the floating pelagic purple snail *Janthina* spp. and the floating pelagic nudibranch *Glaucus atlanticus.* See ***Glaucus atlanticus*** and ***Porpita porpita***.

**cacker.** An undersized rock lobster. Cacker, cakker and kakka are interchangeable. See **rock lobster**.

**caisson.** A structure designed for use underwater for engineering purposes. It consists of an airtight chamber with an open bottom into which air is pumped at high pressure to keep water out. Caissons are used underwater in rivers and harbours in the building of bridge and wharf pylons as well as other structures. The people working in these caissons in earlier times were returned to the surface without any proper recompression and as a result many of them showed the classic symptoms of decompression sickness. See **decompression sickness**.

**caisson disease.** Another name for decompression sickness. See **decompression sickness**.

**cakker.** See **cacker**.

**calipee.** A yellow-green cartilaginous substance cut from the lower carapace of turtles for use in the manufacture of turtle soup. A turtle weighing 100 kg may yield only a few kilograms of calipee. See **carapace** and **turtle**.

**Calypso.** A ship owned by The Cousteau Society and made famous by the underwater explorer Jacques-Yves Cousteau and his team of divers and scientists who are involved in marine research in many oceans of the world. Built as a mine sweeper in the United States in 1942, the *Calypso* is 43 metres long and 7.17 metres in beam with a double hull made of wood. The ship has been modified many times to make it suitable for research work. Its maiden voyage was in 1951 to the Red Sea. Since then it has made many ocean voyages including a research trip penetrating the upper reaches of the Amazon River in South America. See ***Alcyone***.

**Calypso-Phot.** The first commercially available self-contained underwater 35 mm camera—designed by Jacques-Yves Cousteau and Jean de Wouters in 1955 and released for sale by Spirotechnique of France in 1960. The camera was named after Cousteau's research ship *Calypso.* In 1963 the production of the camera was taken over by Nikon of Japan and the camera renamed Nikonos. The Calypso-Phot camera was sold in Australia until at least 1964. See **Nikonos camera**.

**capillary.** A small bore tube; also the smallest blood vessels connecting arteries and veins.

**carapace.** A hard, chitinous shield, or bony outer external covering of certain animals such as crabs and turtles.

**carbon monoxide poisoning.** Contamination of scuba cylinder air supply usually caused by the exhaust fumes of a petrol driven compressor when refilling scuba cylinders. If the compressor is poorly maintained, oil may be ignited

Versatile Amphibious

**CALYPSO-PHOT**

**TESTED TO A DEPTH OF 200 FEET**

35mm. Camera

Waterproof ● Corrosionproof ● Sandproof ● Mudproof

★ Takes Black and White or Colour, 35mm.
★ Interchangeable lens, telephoto, ultra-wire.
★ Spring-loaded, rapid wind and release.
★ Integral flash contact socket.
★ Camera size, 3½ x 2½ x 1¾ inches. Wgt. 25 oz.
★ Focal plane shutter from 1/30 to 1/1,000 sec.
★ Automatic depth of field indicators.

Available all leading camera and sports stores.
Trade enquiries or further details,

**Distributors:**
A.F.C.A. Pty. Ltd.
12 Castlereagh Street, Sydney. 28-2076
Australian Divers Company Pty. Ltd.
506 Spencer Street, West Melbourne. 30-4040

AUSTRALIAN SKINDIVERS MAGAZINE — Page 12 — JANUARY, 19

**This advertisement for a Calypso-Phot camera appeared in an issue of *Australian Skindivers Magazine* in 1964. Special thanks to the Australian Underwater Federation for permission to reproduce this photograph.**

in the cylinder heads, causing carbon monoxide to be produced. Breathing of this contaminated air may result in unconsciousness without warning. The tell-tale sign of carbon monoxide poisoning in an unconscious diver is a bright pink colouration of the lips. Oxygen is the best first aid measure. Air that has an oily odour or taste may be contaminated with carbon monoxide and should not be used. See **compressor** and **diver's air test kit**.

***Catostylus mosaicus.*** Jellyfish and member of the class Scyphozoa. Arguably Australia's most common large jellyfish, growing to about 35 cm in diameter and recognised by its large brownish dome (sometimes white or blue) with eight short arms. It is a common sight in our estuaries and along our eastern coastline, being found in waters off Papua New Guinea and the Philippines in the north, to Port Phillip Bay, Victoria in the south. These jellyfish are harvested in Australian waters for processing and exported as food to Asia. See **jellyfish**.

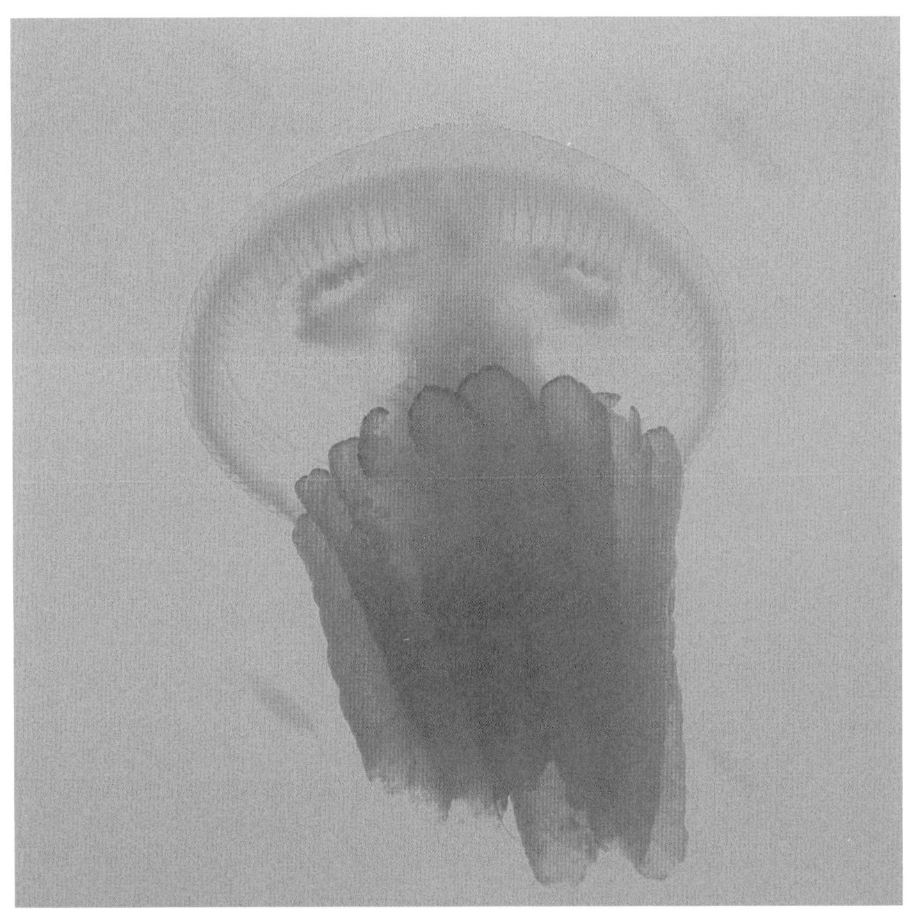

**The large jellyfish *Catostylus mosacicus* is at times so numerous in estuaries on the east coast of Australia it appears to carpet the water.**

**Caulerpa.** A genus of green algae (division Chlorophyta), easily recognised by the creeping stems attached by fine branching filaments to the substrate and bearing erect fronds of varied shapes: leaf-like, grape-like, cylindrical or serrated. In the Philippines *Caulerpa racemosa* is cultivated for use as a salad vegetable. See **algae**.

**The green alga *Caulerpa cactoides* is common along the coastline of southern Australia.**

**The green alga *Caulerpa filiformis* is seasonally very common in shallow water and on the marine rock platforms around Sydney.**

**Cave Divers Association of Australia (CDAA).** An association formed in 1973 which sets out a series of criteria and testing procedures to ensure scuba divers are adequately prepared, trained and equipped for this specialised form of diving. The CDAA produce a quarterly journal titled *GUIDELINES.* For more information contact: CDAA, PO Box 290, North Adelaide, SA 5006.

**cave diving.** Scuba diving in caves. Before attempting cave diving it is important to be trained in this specialised field as many unnecessary deaths have occurred in divers lacking the basic skills. See **Cave Divers Association of Australia**.

**cay.** See **coral cay**.

**'C' card.** A card certifying that its holder has received basic training in scuba diving techniques.

A selection of scuba divers' C cards.

**cephalopod.** Member of the phylum Mollusca, class Cephalopoda. Highly developed molluscs including cuttlefish, octopus and squid which are all active predators feeding mainly on crustaceans and fish. Cephalpods can be further classified depending on the number of gills: subclass Nautiloidea —four gills, the only living members of this subclass are species of the genus *Nautilus*; and the subclass Coleoidea—two gills, the cuttlefish, octopus and squid. See **cuttlefish**, **nautilus**, **octopus** and **paper nautilus**.

**cetacean.** Member of the order Cetacea. Dolphins, whales and porpoises are collectively known as cetaceans and are divided into two sub-orders: the Odontoceti, or toothed whales and the Mysticeti, or baleen whales. See **whale**, **dolphin**, **humpback whale** and **orca**.

**Cetacean Society International.** A conservation group that has joined other conservation organisations to help stop the slaughter of dolphins by commercial tuna fishermen off the California coast in the eastern Pacific Ocean. Contact: Cetacean Society International, PO BOX 9145, Wethersfield, CT 06109, USA.

**channel.** The deeper part of a waterway usually marked with navigational aids.

**Charles' Law.** A law formulated by the French scientist Jacques Charles which states that: *If the pressure remains constant, the volume of a gas will change inversely with the absolute temperature.* For example when a scuba cylinder is filled hot (not placed in container of water), then the internal pressure will decrease when the cylinder cools. Conversely a full cylinder should never be left in the sun as the heat will cause the pressure to increase which may result in the cylinder burst disc exploding, in extreme cases, where no burst disc is fitted or the disc is faulty, the cylinder itself may explode. See **burst disc**.

**Chelicerata.** Phylum of the Animal Kingdom containing the horseshoe crabs and the sea-spiders. See **sea-spiders** and **Xiphosura**.

**chokes.** A serious form of decompression sickness with lung involvement. This condition results from the accumulation of bubbles of nitrogen gas in the lung tissue which causes pain, shortness of breath, coughing and unconsciousness. See **decompression sickness**.

**Christmass tree worm.** See **polychaetes**.

**cichlid.** Member of the Cichilidae family of fishes. There are about 650 known species, most of these are freshwater species but some live in brackish water. Cichlids are commonly sold as aquarium fishes.

**ciguatera poisoning.** The commonest type of fish poisoning in the Pacific region and a continual problem for islanders who eat large numbers of pelagic and reef fishes. The classic symptoms are nausea, diarrhoea, cramps, severe headaches and a strange temperature reversal effect on the senses. Poisoning occurs when infected fish are eaten and is probably caused by the single-celled dinoflagellate *Gambierodiscus toxicus,* which is ingested by small herbivorous fishes, which are in turn eaten by larger carnivorous fishes, this having the effect of concentrating the toxin as it passes up the food chain. Ciguatoxin is one of the most potent marine poisons known to the human species and is concentrated in various organs and muscle tissue of many different types of reef fishes including: surgeon fish, parrot fish, bass, snapper, wrasses, Spanish mackerel, barracuda, moray eels and grouper. In all, more than 400 species of fishes have been incriminated at various times. Particular species of fish vary in their toxicity from reef to reef and with the seasons. Some scientists say that reef disturbance caused by cyclone damage, dynamiting, dredging and filling appears to have some relationship to the appearance of ciguatera-infected fishes. More research is needed in this area. An article, in a 1988 issue of *The Journal of the American Medical Association,* reported that doctors from the Marshall islands and the University of Hawaii have been experimenting with the use of a simple sugar, mannitol, in the treatment of ciguatera poisoning, with some success. This treatment brought severely poisoned patients out of comas and relieved other debilitating symptoms within minutes without causing any complications. See **saxitoxin**.

Spanish mackerel from certain areas off the Queensland coast are considered unsafe for eating as they can at times cause ciguatera poisoning.

**clam.** A bivalve mollusc and member of the family Tridacnidae. Clams have an exposed mantle, in which they farm symbiotic dinoflagellate algae called zooxanthellae. This symbiotic relationship provides food and protection to the algae while the clam obtains nutrients in the form of by-products from the metabolism of the algae. Clams may be free living or burrowing.

There are two common free living clams on the Great Barrier Reef:

• *Tridacna gigas,* the giant clam, which can attain a metre or more in length and a weight of 200 kg. It was once believed that these giants had the power to hold a person underwater and thereby cause drowning. Researchers have observed that clams close their shells slowly, so entrapment of human limbs is extremely unlikely.

**The free-living giant clam *Tridacna gigas.***

• *Hippopus hippopus,* the horseshoe clam, can grow to 30 cm in length. When walking over coral reef flats at low tide these clams can sense movement nearby and contract their abductor muscles, causing the shell to close and a jet of water to shoot into the air.

The two common burrowing clams on the Great Barrier Reef are:

• *Tridacna crocea,* a small species growing to 15 cm in length, which burrows so deeply into the coral substrate that only its colourful fleshy mantle is visible.

• *Tridacna maxima,* a larger species growing to 30 cm in length which is generally found loosely embedded in the coral. See **symbiosis** and **zooxanthellae**.

The burrowing clam *Tridacna maxima.*

**clearing the ears.** See **Valsalva manoeuvre**.

**closed circuit breathing apparatus.** An underwater breathing device in which the expired gas is recycled by removing the excess carbon dioxide and replacing it with fresh oxygen from a small cylinder. These devices, which do not leave obvious bubble trails on the surface, are also called oxygen rebreathers and were used extensively by various navies during the Second World War for underwater attacks on moored ships and for mine clearance. The main disadvantage of closed circuit equipment is its complexity and requirement for frequent and expert maintenance. Divers using this equipment can lapse into unconsciousness without warning at depths greater than 10 metres due to oxygen toxicity, and a water leak may produce a lethal gas from the expired gas filtration/scrubbing compound. This equipment is not recommended for use by sports divers. See **oxygen toxicity**.

**clouds as weather indicators.** Cloud formations that can signal impending weather conditions:
- Altocumulus clouds generally indicate continued good weather.
- Clouds in layers and at different levels indicate changeable weather.
- Gathering altostratus cloud means an approaching warm front and rain.

**CMAS.** See **Confederation Mondiale des Activities Subaquatiques**.

**coastal currents.** Non-tidal currents caused by oceanic circulation and weather.

**Coast and Wetlands Society Inc.** The following information has been supplied by the Coast and Wetlands Society:

A Society with the following aims and objectives—

- To promote the appreciation and conservation of all aspects of coast and wetland ecosystems.
- To promote the application of ecological principles in the conservation, development and utilisation of coastal and wetland ecosystems.
- To advise government and other agencies, where the Society may be of assistance.
- To conduct research into aspects of coastal wetland ecology.
- To publish results of scientific investigations and other material designed to encourage conservation and appreciation of coastal and wetland ecosystems.
- To increase public knowledge and awareness of aspects of coastal and wetland ecology and conservation.

Members of the Society receive a regular newsletter and its journal *Wetlands*. Membership affords the opportunity to be actively involved in conservation of coastal and wetland resources, through research and educational programmes. Contact: The Coast and Wetlands Society, PO Box A225, Sydney South, NSW 2000.

**coelenterate.** A member of the phylum Coelenterata. A group of invertebrate animals having a saclike body and central stomach cavity, found living in the water column, on the sea bottom and on other marine organisms from the shore to abyssal depths, they include; corals, hydroids, jellyfishes, sea-anemones and zoanthids.

**Coleman, Neville.** Well known Australian underwater photographer and naturalist who has written a number of marine identification guides and is the editor and publisher of the *Underwater Geographic* journal. See **Australasian Marine Photographic Index** and ***Underwater Geographic***.

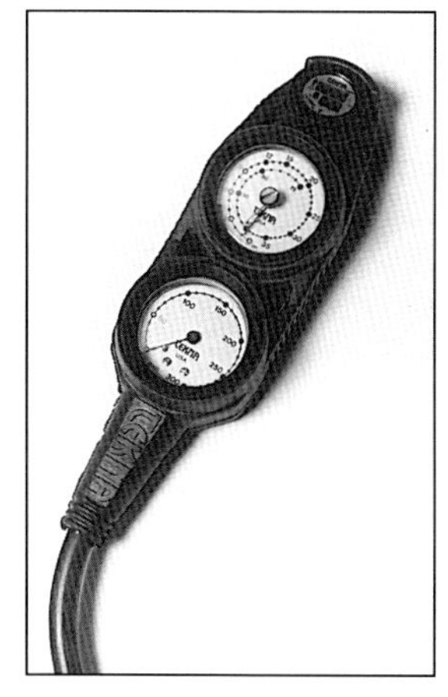

**The combination or combo gauge consists of a number of diving instruments in a single cluster which is connected to a high pressure outlet on the first stage of a regulator.**

**combination gauge/combined dive gauge.** A number of diving instruments combined in one unit. Also known as a 'combo' gauge. Combination gauges have as a minimum a contents and depth gauge, but may also incorporate a compass, bottom timer, watch, thermometer and/or dive computer.

**comb jelly.** See **ctenophore**.

**combo.** See **combination gauge**.

**commercial diving.** All diving operations, other than scientific diving or search and rescue diving, carried out as part of a business or for profit.

**compass.** An instrument that has a magnetised freely moving needle to indicate magnetic north, especially useful for diving at night, on featureless bottoms or in dirty water. The recommended type of underwater compass should be oil filled with luminous dial and needle, and a rotating bezel. Practice is required to master underwater navigation as it is not easy to swim a set course using a compass alone to find your way.

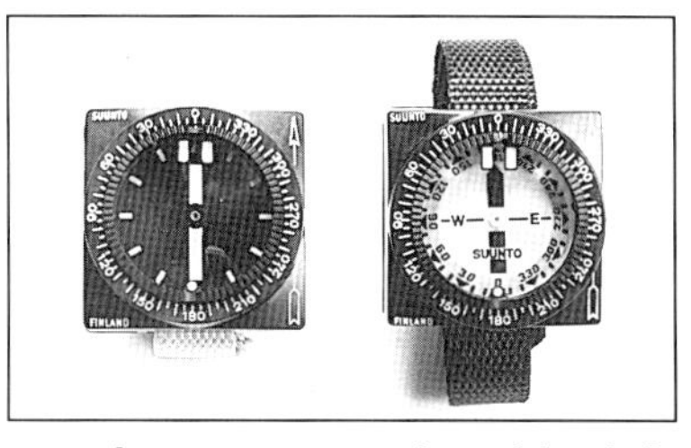


An underwater compass is useful to help determine your direction of travel on a featureless bottom or in murky water.

**compound ascidian.** Member of the subphylum Urochordata, class Ascidiacea. Compound ascidians are comprised of large numbers of individual animals living together and embedded in a common jelly-like matrix. Compound ascidians are in most cases found encrusting on all manner of things from marine invertebrates and plants to rocks and man-made objects. See **ascidian** and **tunicate**.

The compound ascidian *Botrylloides* sp. These ascidians have a wide geographic range from the tropics to the waters off southern Australia.

**compressor.** A petrol or electric driven motor which powers a compressor unit which is designed to compress air for filling scuba cylinders. There is a large number of brands and capacities available the larger models can fill a standard 80 cubic foot scuba cylinder in as little as a few minutes but the smaller models may take from 20 to 60 minutes to fill the same size cylinder. The smaller compressors up to about 4 cfm in capacity can be transported easily for use by scuba divers at isolated locations. The larger and much heavier models are used at dive stores and on dive charter vessels. See **diver's air test kit** and **hookah**.

**conchologist.** A person who collects and studies molluscs (shells). Also called a malacologist.

**cone shell.** Mollusc and member of the family Conidae. There are over 500 species of cone shells described worldwide, a small number of which are considered extremely dangerous to humans including: *Conus geographus* (several deaths attributed to this species), *C. aulicus, C. catus, C. magus, C. marmoreus, C. monachus, C. omaria, C. striatus, C. textile* and *C. tulipa.* Cone shells belonging to the species listed should not be handled. If a cone shell is picked up, the proboscis can be extended and the specialised radula fired like a harpoon, injecting venom which can be very painful and in some cases fatal. A cone shell can inject venom even through clothing and so should not be placed inside a pocket. If you have to handle cone shells, pick them up at the wide end of the shell using thick gloves. Even an apparently empty shell of any kind may contain a blue-ringed octopus so scuba divers should never store shells in pockets or wetsuits. Some cone shells are highly prized by collectors for their beauty and rarity, including; *C. gloriamaris* the famous 'glory of the seas' cone found in the waters off Indonesia, the Philippines and Fiji. See **proboscis**.

**The cone shell *Conus marmoreus* is considered dangerous to humans and is common throughout the Great Barrier Reef and the Indo-Pacific region.**

**Confederation Mondiale Des Activities Subaquatiques (CMAS).** The following information has been supplied by CMAS:

The Confederation Mondiale des Activities Subaquatiques (CMAS) or World Underwater Federation, was formed at Monaco in 1959, by 15 countries, under the care of Jacques-Yves Cousteau. Today this organisation comprises national federations and associations in 73 countries and millions of active divers.

CMAS has established a system of international certification that allows the standardisation of diving instruction around the world. The true United Nations of Diving, CMAS has established itself as the sole organisation representing the great international community of divers. Its activities, whether of a sporting, technical, scientific, medical or cultural nature, are open to everyone without any racial, social, political or religious discrimination.

As an independent, non-profit making organization, CMAS sets itself the objective of the harmonious development of underwater activities and the coordination of diving instruction around the world. CMAS, 47 rue du Commerce, 75015 Paris France; phone: 45.75.42.75, telex: 205 734 F, fax: 45.77.11.04.

**contents gauge.** A measuring device attached to the high pressure outlet on the first stage of a regulator to indicate the pressure of the air remaining in a scuba cylinder. A reading of zero indicates ambient pressure, so no more air can be breathed from a cylinder at this pressure. Contents gauges are normally purchased as a single unit or in combination with a depth gauge and/or a compass, generally referred to as a combo gauge.

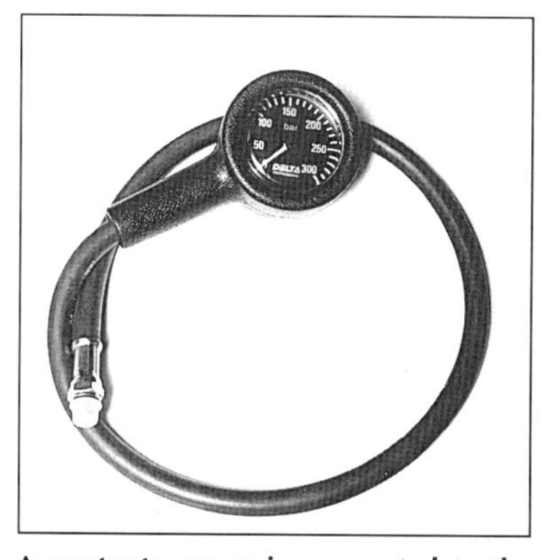

A contents gauge is connected to the first stage of a scuba regulator so that the quantity of air remaining in the cylinder can be monitored throughout the dive.

**continental shelf.** The region of the sea floor that extends from the low water mark down to 200 metres (656 feet) depth around each continent.

**continental slope.** The steep underwater slope that extends from the continental shelf seaward to the deeper oceanic regions.

**cookiecutter shark.** Member of the order Squaliformes, family Squalidae, also known as the cigar shark or the luminous shark (because of light-producing organs on the ventral surface of the body and fins). This small deep-water shark grows to about 50 cm in length and has unusual feeding habits, obtaining nourishment by biting biscuit-sized pieces from the bodies of whales, sharks, dolphins, squid and large fish. Cookiecutter sharks are normally found in water deeper than 500 metres but are sometimes captured near the surface at night. There are two known species, *Isistius brasiliensis* and the largetooth variety *Isistius plutodus.*

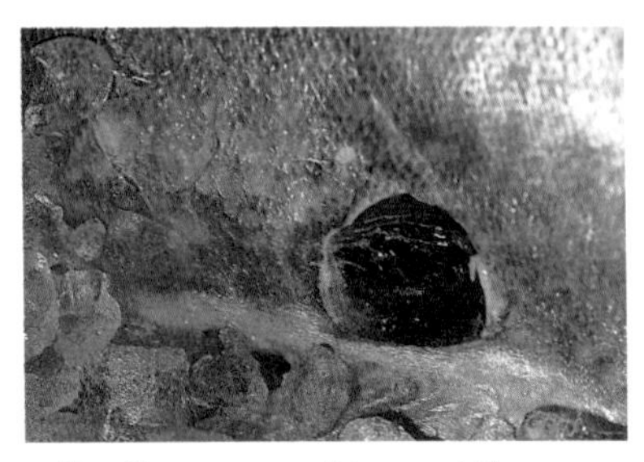

**Yellowfin tuna caught on set lines over deepwater are sometimes attacked by cookiecutter sharks. These sharks do not kill the tuna but leave characteristic small circular bite marks on the fish.**

**coral.** The common name for the order Scleractinia, all members of which have hard limestone skeletons. There are reef-building (hermatypic) and non-reef building (ahermatypic) groups of corals. The former need shallow sunlit waters to live and the latter can live in twilight conditions in deeper water. See **hard coral** and **soft coral**.

**coral bleaching.** Extremely pale coloured or pure white coral due to the loss of symbiotic algae by the coral polyps. Bleaching of hard corals occurs as a result of stress, and is usually observed in the summer after periods of high temperature or heavy rainfall. In severe cases the bleached corals will eventually die; in less severe cases, they will regain their symbiotic algae and make a full recovery. Coral bleaching has occured on the Great Barrier Reef and on some coral reefs off the Western Australian coast. The Great Barrier Reef Marine Park Authority is anxious to determine the severity and extent of bleaching and has issued forms for reporting bleaching. The forms are available from: Great Barrier Reef Marine Park Authority, PO Box 1379, Townsville, Qld 4810 Fax: (007) 72 6093. See **hard coral** and **zooxanthellae**.

**coral cay.** Small islands which have formed on coral reefs from sedimentary debris washed up on the reef flat. Initially made from coral sand and rubble, a cay becomes more stable as calcium carbonate precipitates out of the ground waters of the cay and cements together underground sand and rubble to form a cement-like substance called beach rock. This rock forms an effective barrier against erosion. Coral cays are known as 'keys' in other parts of the world.

**A coral cay on the southern section of the Great Barrier Reef complex.**

**coral crayfish.** See **tropical rock lobster**.

**coralline alga.** A member of the division Rhodophyta (red algae). Like corals, coralline algae contain calcium carbonate. Australian coasts are rich in coralline algae. They grow on many different substrates—rocks, shells, ascidians, other algae and sea-grasses. Coralline algae may be encrusting or erect with flattened articulated branches. In colour they range from bright pink to red and purple.

**Coralline algae can grow as an encrusting layer over corals, rocks, shells, and other algae.**

***Corallina* sp. is an erect feather-like form of coralline algae.**

**coral reef.** Limestone reefs built by living plants and animals. The main reef builders are coral polyps and coralline algae. Over thousands of years the calcium carbonate skeletons of marine plants and animals are built-up and compacted into a reef many metres thick. The reef has a veneer of living corals and a host of associated plants and animals which add to the building process year by year. For example, parrot fish eat hard coral, digesting the polyps and zooxanthellae; the calcium carbonate passes through the digestive tract and the combined efforts of millions of parrot fish adds thousands of tonnes of coral sand to coral reefs each year. The Great Barrier Reef comprises the largest group of coral reefs in the world.

**A typical coral reef scene in a lagoon protected from pounding surf where erect branching forms of staghorn corals (*Acropora* spp.) can predominate.**

**coral spawning.** The release of eggs and sperm by corals. Coral spawn is sometimes in a package large enough to be seen with the naked eye. A mass spawning takes place every year on the Great Barrier Reef for a few nights of several consecutive months in late spring and early summer. Coral spawning is a synchronised event of spectacular proportions. Once expelled from the oral cavity of the coral polyps, the spawn bundles float to the surface and break apart where fertilisation takes place within a few hours. The developing embryos may be swept by ocean tides and currents to new areas where they settle onto a suitable substrate and grow into new coral colonies.

**core temperature.** The internal temperature of deep body tissue. The human core temperature is normally 37°C and when this temperature falls a person is said to be suffering from hypothermia. See **hypothermia**.

**Corlieu, Louis de.** The inventor of the first rubber swimming fins in 1929. He patented his invention in France in 1933.

**Cousteau, Jacques-Yves.** Famous scientist, underwater explorer and inventor, born in 1910 in St. Andre-de-Cubzac, France. Cousteau and Emile Gagnan jointly developed the scuba demand value in 1942 which paved the way for the development of the modern scuba regulator and opened up the underwater world for sport diving. Cousteau and his co-workers began the science of underwater archeology, made the first ocean floor search for and discovery of petroleum, invented many new tools for underwater research and filming, created the first small submarine for scientific work, carried out the first successful experiments in living under the sea and invented the first underwater television system. Cousteau has witnessed over the past three decades the rapid destruction of our planet, the pollution of our oceans, rivers and lakes. Under his guidance The Cousteau Society is educating people and nations to become more environmentally aware. See **Calypso**, **Cousteau Society** and **Gagnan**.

**Cousteau Society.** The following information has been supplied by The Cousteau Society:

A non-profit, membership-supported organisation dedicated to the protection and improvement of the quality of life. The Society believes that an informed and alert public can best make the choices to provide a healthier and more productive way of life for itself and for future generations. To this end, The Society conducts an array of investigative programs and carries out research in fields that have little chance of being funded by government or industry.

Under the leadership of Jacques-Yves Cousteau and Jean-Michel Cousteau, The Society's award winning television specials serve to inspire interest in the marine environment, and to foster a concern throughout the world for marine life and for the quality of all life.

During three decades, The Cousteau Society's research vessel *Calypso* has sailed on major expeditions worldwide, from the Antarctic to the Red Sea, from the Mediterranean to the Pacific and the remote jungles of the Amazon.

Uniquely equipped as a floating research station, laboratory, diving platform, and photographic facility, *Calypso* travels free of obligations to nations or corporations.

The Society remains independent because of the individual contributions of Cousteau Society members. In search of desperately needed understanding of changes taking place in the planet's water system, and to help determine solutions to environmental problems, The Cousteau Society's teams travel to wherever the problems are, to observe close-up the life processes of ocean and river systems.

They analyse, then document on film, vital areas where human activities are causing changes in the living patterns of whales, dolphins, fishes, and other aquatic creatures, as well as the human communities which may be dependent upon them. The Cousteau Society, 930 West 21st Street, Norfolk, Virginia 23517, USA.

**cowrie.** Marine gastropod mollusc and member of the family Cypraeidae. The colourful and highly polished cowries are one of the most popular shells with collectors. One of the rarer cowries *Cypraea aurantium* or the 'golden cowrie' has a limited range from the Philippines and New Guinea in the western Pacific to Polynesia and is prized by collectors.

**These tiger cowries, *Cyprea tigris*, were found under the small ledge in the background.**

**coxswain.** The person who steers a boat. Also called a helmsman.

**CPR.** Cardio Pulmonary Resuscitation. A technique of alternating between External Cardiac Compression (ECC) and Expired Air Resuscitation (EAR) or mouth to mouth/nose as it is commonly called. The CPR technique is used, when both the patients heart and breathing have stopped.

**crack.** A term used to describe briefly opening and closing the valve on a compressed air cylinder.

**cray.** See **rock lobster**, **tropical rock lobster** and **western rock lobster**.

**crayfish.** See **rock lobster**, **tropical rock lobster** and **western rock lobster**.

**crinoid.** See **feather-star**.

**Crop, Ben.** Australian underwater cinematographer who produced a number of TV specials depecting marine life off the Australian coast. He has also established a shipwreck treasure museum in Port Douglas, north Queensland.

**crown-of-thorns starfish.** Echinoderm and member of the class Asteroidea. The crown-of-thorns starfish *Acanthaster planci* is found in tropical Indo-Pacific waters where it has caused extensive damage to corals. The Great Barrier Reef (GBR) has been extensively damaged by the crown-of-thorns in recent years but some experts believe the increase in numbers of this starfish is a natural cycle, although this theory has not been scientifically proven one way or the other. The Great Barrier Reef Marine Park Authority (GBRMPA) received a grant from the Commonwealth Government to coordinate a research programme on the starfish and scuba divers can help in this research by reporting the presence or absence of the crown-of-thorns next time they visit any part of the GBR. Special report forms are available from the Great Barrier Reef Marine Park Authority. A survey carried out by the Australian Institute of Marine

This crown-of-thorns starfish was turned over to reveal its stomach which was everted to directly absorb the coral polyps on which it feeds.

This unusual purple coloured variant of a crown-of-thorns starfish was photographed in the Similan Island group in Thailand.

Science (AIMS) between November 1989 and January 1990, reported only two starfish found on 45 reefs surveyed between Lizard Island and Innisfail. This may indicate that the crown-of-thorns plague could be a cyclic natural phenomenon. GBRMPA has produced a booklet and videotape titled 'The Crown-Of-Thorns Story' which is available from: GBRMPA, PO Box 1379 Townsville, Qld 4810; or from: Great Barrier Reef Wonderland Aquarium, PO Box 1555, Townsville Qld 4810.

**The crown-of-thorns starfish is a well known coral feeder and has caused a great deal of damage on the Great Barrier Reef in recent years. The freshly killed coral has a white bleached appearance.**

**ctenophore.** A member of the phylum Ctenophora, commonly called a comb jelly or sea-gooseberry. Ctenophores have delicate transparent bodies and are found in the surface plankton of all seas. They are noted for their green-blue luminescence (seen at night) which originates from beneath the comb rows (transverse rows of fused cilia called combs). Ctenophores are hermaphroditic

**This fragile ctenophore was photographed at Heron Island, Queensland; the diver gives an indication of its size.**

and active predators, feeding on plankton which they catch with sticky cells that act like flypaper. They do not have stinging cells (nematocysts). See **hermaphrodite** and **plankton**.

**cunjevoi.** See **tunicate**.

**current.** A tidal or non-tidal flowing body of water. Tidal currents are due to the gravitational forces of the Sun and Moon on the Earth which causes the oceans to rise and fall at regular intervals. Non-tidal currents such as the East Australian Current are part of the ocean's circulatory system. Storms and changing wind patterns can effect both tidal and non-tidal currents.

**cuttlefish.** Cephalopod mollusc and member of the order Sepioidea. The cuttlefish are extremely advanced, intelligent and mobile molluscs which have an internal shell or cuttlebone that gives strength and rigidity to their body.

**The giant cuttlefish *Sepia apama* is easily approached by scuba divers.**

Cuttlebone is made from calcium carbonate and contains a gas which provides buoyancy. When cuttlefish die their soft bodies decompose but the hard skeleton of cuttlebone remains and floats on the surface until washed ashore where it is often collected by bird fanciers to supplement their pets' diet. Cuttlefish have 10 tentacles attached to the head, two of which are longer than the rest and have sucker pads with spoon shaped ends which are used for capturing prey and transferring it to the mouth parts. The mouth consists of a chitinous beak, in the centre of the group of tentacles. A continuous fin along each side stabilises the cuttlefish and aids in propulsion. When full speed is required the tentacles are brought together rapidly for streamlining, and water is expelled from the siphon for propulsion. See **cephalopod**.

When cuttlefish die, a buoyant gas filled cuttlebone remains. Centuries ago cuttlebone was crushed and used as a crude form of toothpaste. Today it is collected by bird fanciers for their pets to eat.

**Cyalume.** A brand name for a chemical lightstick. The device consists of a clear plastic tube containing separate compartments for two chemicals. When the tube is bent the chemicals are mixed together to cause a chemical reaction which results in a yellow-green light lasting for several hours. Cyalumes are commonly used by night divers, not to see by, but as markers on their equipment, to enable other divers or rescuers to find them if their main dive light fails. Lightsticks are now available in a number of different colours. See **night diving**.

**cyanobacteria.** A member of the division Cyanophyta, previously known as blue-green algae. *Rivularia firma* is a common species on Australia's southern rocky coasts where it forms round dark-green gelatinous colonies in the barnacle zone. Certain species of cyanobacteria inhabit living sponge tissue, e.g. *Oscillatoria symbiotica* lives in the tissues of the tropical marine sponge *Dysidea herbacea*.

The cyanobacterium *Rivularia firma* encrusts rocks on our southern coastline.

**cyclone.** An atmospheric depression, characterised by clockwise motion of winds (up to 240 km/hr) in the southern hemisphere and anticlockwise wind movements in the northern hemisphere. Cyclones generally stay over the sea but occasionally move onto the coast where they sometimes cause great devastation, as occurred on Christmas Day 1974 when the city of Darwin in the Northern Territory was destroyed by cyclone *Tracy*.

**Dalton's law.** The General Gas Law formulated by the English scientist John Dalton which states that: 1) *The total pressure a mixture of gases exerts is equal to the sum of the separate pressures which each of the gases would exert if it alone occupied the whole volume.* 2) *The partial pressure is the pressure exerted by a constituent gas that occupies the whole volume occupied by the mixture at the same temperature and pressure in the absence of the other gases.* To understand the physiology of scuba diving it is necessary to be able to explain and predict how component gases within a mixture will behave when placed under pressure.

**DCIEM Sport Diving Tables.** A set of air decompression tables developed at the Canadian Defence and Civil Institute of Environmental Medicine (DCIEM) in Toronto, Canada. The sport diving edition of these tables was released in October 1987 and is the result of nearly three decades of research. The new tables are designed for recreational scuba diving and include instructions for multi-level diving, high altitude diving, minimum surface interval calculation, omitted decompression and flying after diving. The new tables appear to be much more conservative than the United States Navy tables. The DCIEM tables, guides and manuals are distributed in Australia through the Royal Adelaide Hospital, Hyperbaric Medicine Unit, North Terrace, Adelaide, South Australia 5000; phone: (08) 223 0230. See **air decompression tables**.

**decanting.** Transfer of air from a cylinder of high pressure to one of a lower pressure. This occurs when filling scuba cylinders from an air bank. See **air bank**.

**decompression chamber.** See **recompression chamber**.

**decompression dive.** A dive that exceeds the 'No-Decompression' time limits given in the dive tables and that requires 'Decompression Stops' at specified depths for the elimination of excess nitrogen in the diver's tissues. Decompression dives should be avoided if at all possible. The following equipment should be available for decompression dives:
- A descent/ascent line suspended from the boat.
- A spare scuba cylinder and regulator on a weighted shot line at the first scheduled decompression stop.
- A waterproof set of air decompression tables. All calculations should be done before the dive.
- The largest capacity scuba cylinder available should be used.
- An octopus demand valve should be used by each member of the dive team.
- All divers should have a BC, watch, contents gauge and depth gauge. A bottom timer and compass are useful.
- Oxygen should be available on the boat.
- The boat person should have knowledge of the nearest recompression chamber and other emergency services such as the DES.
- The boat should contain a marine radio transceiver.

See **air decompression tables** and **Diving Emergency Service (DES)**.

**decompression meter.** A device that automatically takes account of a diver's time at depth and suggests a decompression schedule and/or surface interval before the next dive. In the past these meters have been unreliable, although the new type of electronic computer meters may prove to be more reliable. Seek advice from an un-biased source e.g. your local diving doctor, before buying one. See **dive computer**.

**decompression sickness.** Also known as 'the bends'. A condition resulting from the formation of bubbles in the blood or tissues following or during ascent or decompression. Symptoms include pain in or around a joint, an itch or rash and localised swelling. If any of these symptoms are evident after scuba diving it is important to seek medical assistance by phoning the Diving Emergency Service (DES) on **(008) 088 200**. Physiological and environmental factors can increase the chances of developing this ailment. Factors which increase the risk of decompression sickness include: exertion, cold water, consumption of alcohol before or immediately after a dive, dehydration, obesity, repetitive dives, age and a previous history of decompression sickness. Factors decreasing the risk of decompression sickness include: physical fitness, ascending slowly, building up to deeper dives by doing a series of progressively deeper dives, doing a safety stop at 3-5 metres on every dive, and on repetitive dives, doing the deepest dive first. See **Diving Emergency Service (DES)** and **recompression chamber**.

**decompression stop.** A scheduled interruption of the diver's ascent at a specified depth for a specified time, to allow the elimination of excess nitrogen from the diver's body. See **air decompression tables**.

**decompression tables.** See **air decompression tables**.

**decompression time.** The time spent decompressing, this includes the ascent time to the decompression stop. See **air decompression tables**.

**deep diving records.** See **breathhold** and **scuba deep diving records**.

**deepest manned ocean descent.** The bathyscaphe *Trieste* manned by Jacques Piccard and Lt. Donald Walsh descended to a depth of 10916 metres in the area of the Marianas Trench, 400 km southwest of Guam in the Pacific Ocean on the 23 January 1960.

**demand valve.** The part of a scuba regulator which reduces the line pressure in the hose coming from the first stage (part attached to cylinder) so air can be breathed at ambient pressure (that of the surrounding water), also called a second stage. The three main types of demand valve are the upstream valve, downstream valve and the servo-assisted valve. See **upstream**, **downstream** and **servo-assisted valves**.

The demand valve or second stage of a scuba regulator is the part placed in the diver's mouth.

**density.** The mass of a substance expressed as weight per unit volume, e.g. kg/cm.$^{3}$

**depth gauge.** A gauge that measures depth below sea-level. There are several types:

• The capillary gauge is the cheapest and most robust and consists of a small diameter plastic tube open to the water at one end and mounted on a calibrated dial. As the diver descends the air in the tube is compressed, causing water to enter the open end. The depth is read on the scale adjacent to the water–air interface in the tube. This gauge is very accurate in water less than nine metres. At depths greater than this the depth calibrations on the gauge are so close together that readings become difficult. Maintenance consists of washing in freshwater and occasionally blowing out salt crystals in the plastic tube.

• Bourdon tube gauges transmit ambient pressure into a curved metal tube which tends to straighten as pressure increases. Linked to a drive gear, the movement of the tube activates the gauge indicator needle from which the depth is read. A modified version called a closed Bourdon tube gauge is also available in which the metal tube is oil filled and capped at one end with a diaphragm. The modified type is supposed to be less susceptible to the effects of corrosion and more sensitive than the open Bourdon tube gauge. Maintenance consists of washing in fresh water to prevent obstruction of the water-entrance port by salt crystals.

• Oil-filled depth gauges feature a compressible outer case that on descent activates a needle-indicator on the calibrated dial face giving the depth reading. Maintenance consists of washing in fresh water.

• Digital depth gauges are the newest on the market. They are complex battery powered electronic instruments and are considered to be the most accurate. See **dive computer** and **personal dive sonar**.

**Depth gauges can be worn on the wrist as a single unit or as part of a combination gauge attached to the first stage of a scuba regulator.**

**depth sounder.** An electronic device consisting of a transducer and recorder used on boats to record depths, presence of schools of fish or solid objects such as shipwrecks, reefs or drop offs. The transducer transmits a signal through the water beneath the boat until it hits a solid object and it is then bounced back to the sounder which converts it to a series of marks on heat sensitive paper or into a graphic display on a TV type monitor which the operator interprets. See **personal dive sonar**.

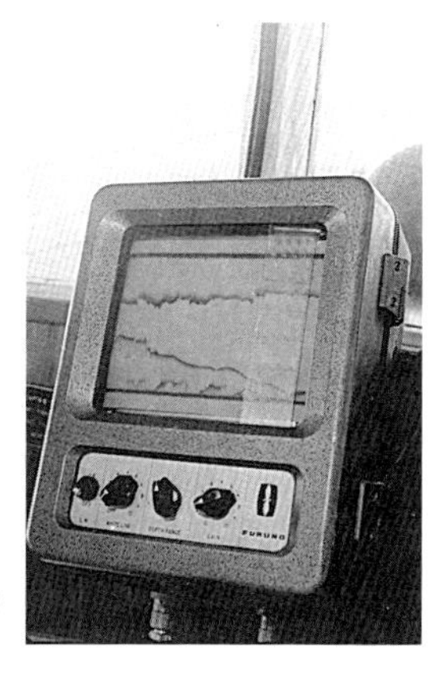

**The depth sounder is an electronic instrument which records depths.**

**derelict.** A ship or object abandoned by the owner at sea. See **flotsam**, **jetsam** and **ligan**.

**DES.** See **Diving Emergency Service**.

**diabetes.** A medical condition characterised by large swings in blood sugar levels due to the lack of insulin secretion by the pancreas. This can cause convulsions and/or loss of consciousness which might be fatal if they occurred underwater. Diabetics can lapse into unconsciousness if their insulin dosages and their food intake are not in balance. It is therefore unwise to dive if you suffer from this condition. See **diving medical examination**.

**diaphragm.**[1] A circular piece of rubber or synthetic material which acts as a partition between the low pressure air in the demand chamber of a regulator and the surrounding water.

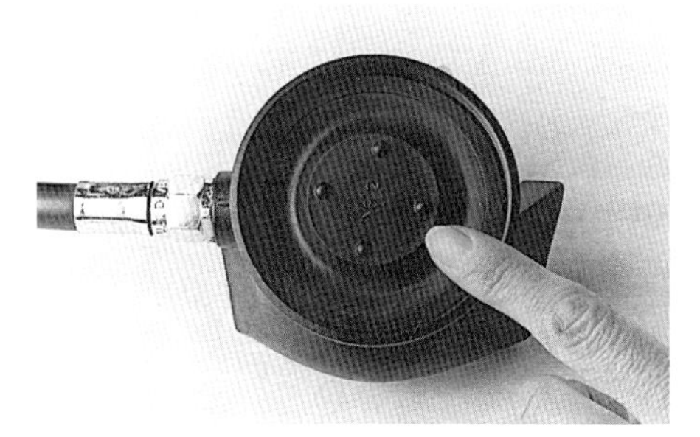

**A silicon rubber diaphragm in place on the second stage of a regulator.**

**The regulator diaphragm has a metal disc attached on the reverse side to prevent damage to the soft rubber by the lever valve mechanism.**

**diaphragm.**[2] In mammals the thin muscle and connective tissue which separates the thoracic cavity from the abdominal cavity.

**dinoflagellate.** Unicellular marine alga and member of the order Dinoflagellida. Dinoflagellates make up part of the marine phytoplankton and are generally a golden brown colour due to the presence of the pigment xanthophyll. These single celled organisms with bizarre shapes, have flagella (for propulsion) and processes (for floating). Several species give off brief flashes of light, others parasitise marine animals, while some are highly toxic to fishes and humans. The zooxanthellae that are symbiotic inhabitants of many corals belong to this group. See **phytoplankton** and **red tide**.

**distress flare.** A basic item of safety equipment that indicates distress in an emergency at sea. There are three main types of distress flares:

- Hand held smoke flares for use in daylight and visible up to 10–15 km in good weather.
- Hand held flares for night use and visible up to 10–20 km in good weather.

• Red parachute flares for use at night and visible up to 40-60 km in good weather. See **boating safety equipment**.

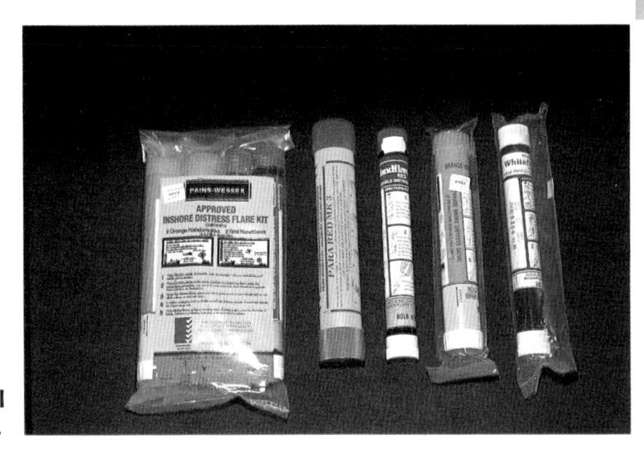


**Distress flares are an essential item of boating safety equipment.**

**dive computer.** An electronic device used by scuba divers which automatically calculates and displays information (via a liquid crystal display) about dive profiles including: depth, elapsed dive time, ascent rate, bottom time, decompression stops/times, surface intervals and a host of other information depending upon the brand of dive computer purchased. The dive computer is helpful when a series of dives are to be undertaken on the same day as no complicated calculations are necessary to determine safe repetitive dive times. You should look for a model which has an ascent rate warning, this feature should be considered as an essential feature. For more information a good reference is *Dive Computers—A consumers guide to History, Theory and Performance* by Ken Loyst, Karl Huggins and Michael Steidley; Watersport Publishing.

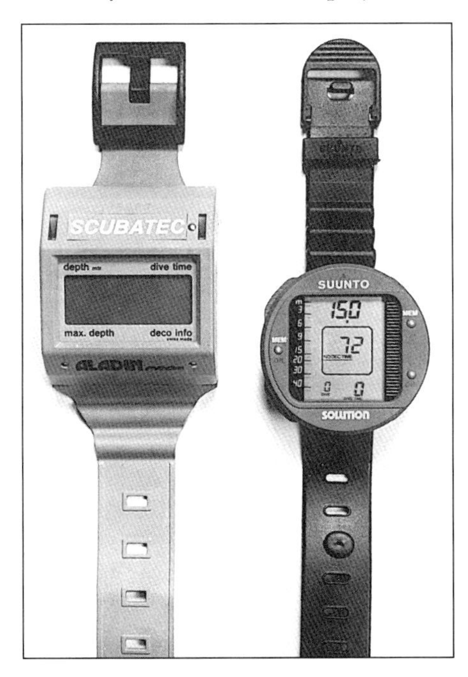


**Modern dive computers**

**dive knife.** A knife usually contained in a sheath of plastic material worn by a diver generally on either the lower outer or inner leg. A dive knife usually has two cutting surfaces a conventional one and a serrated one for heavy use. The conventional smooth cutting edge on new knives is sometimes blunt and may have to be sharpened before use as a blunt knife is useless to a diver entangled in kelp, old fishing lines, nets, or rope. The dive knife is one of the most common items of equipment lost by scuba divers generally because of carelessness and/or bad design. Sheath straps should be made of rubber so they will not slacken at depth when the wetsuit is compressed.

A selection of dive knives.

**dive magazines and newspapers.** Magazines and newspapers produced for divers. The main dive publications currently produced in Australia.
• *Dive Log Australia;* Mountain, Ocean and Travel Publications PO Box 167, Narre Warren, Vic. 3805; phone: (059) 443 774, fax: (059) 444 024.
• *Scuba Diver;* Yaffa Publishing Group, PO Box 606, Sydney, NSW 2001; phone: (02) 281 2333, fax: (02) 281 2750.
• *Sportdiving in Australia & the South Pacific;* Mountain, Ocean and Travel Publications, PO Box 167, Narre Warren, Vic. 3805; phone: (059) 443 774.
• *Underwater Geographic;* Sea Australia Productions, PO Box 702, Springwood, Qld 4127; phone: (07) 341 8931. See ***Asian Diver* magazine**.

**dive mask.** See **face mask**.

**'Diver Below' flag.** The internationally-recognised symbol of a divers presence in the water. Vessels of any kind engaged in diving operations must exhibit the International Code Flag 'A' or a rigid replica of at least 75 cm x 60 cm. Divers not operating from a

The 'Diver Below' flag is instantly recognisable by scuba divers but is not so well known by non-diving boat owners.

vessel must show a smaller version of the flag displayed from a buoy or float which must measure 30 cm x 20 cm and must be 20 cm above the water. These regulations apply to both scuba and breathhold divers. On sighting the 'Diver Below' flag a vessel under way must proceed at a safe speed, keep well clear and maintain a proper look-out for persons in the water until well clear of the area.

**diver propulsion vehicle (DPV).** An underwater vehicle used for towing scuba divers consisting of a small battery powered electric motor with attached propeller enclosed in a waterproof case. Diver propulsion vehicles or DPVs as they are commonly known have been used for over 25 years. The scuba diver holds on to handles at the rear or at the sides to steer the DPV and is carried along underwater by the vehicle. Such a device allows a diver to extend the dive range and bottom time. The large aluminium Farrallon DPVs (powered by up to four 12 volt batteries) as used in the James Bond movie *Thunderball,* have given away to much lighter and smaller plastic models like the Dacor Sea Sprint which has a total weight of 18 kg (including battery), above water and 650 grams underwater. DPVs are a little expensive for the average scuba diver but they can be hired from dive shops, dive charter boat operators, and dive resorts.

**DPVs (diver propulsion vehicles) can extend the range and duration of a scuba dive.**

**diver's air test kit.** A product which is used to test the quality of compressed air in scuba cylinders. The levels of carbon dioxide, carbon monoxide, water and oil vapours can be determined using this test kit. For more information contact: Drager Australia PTY LTD, 134 Moray Street, South Melbourne Vic. 3205; phone: (03) 698 4666.

**Divers Alert Network (DAN).** An American non-profit corporation formed in 1980 to provide a medical insurance scheme for scuba divers, its services include:

- Emergency medical consultation for diving accidents.
- Recompression chamber treatment coordination.
- Medical transportation arrangements for diving victims.
- Training of medical personnel.
- Research and diving accident analysis.
- Diving emergency phone: (919) 684 8111. This service is not available in Australian waters.

• Non-emergency information write to: Duke University Medical Centre, PO Box 3823, Durham, NC 27710, USA or phone: (919) 684 2948.

**Divers Exchange International (DEI).** The following information has been supplied by Divers Exchange International:
A worldwide computerised buddy service for divers. For a small annual charge, members may call or write to DEI to find a buddy or obtain information about specific dive destinations. Membership benefits include:
- Registration in a worldwide network.
- Detailed information on diving services including diving doctor and nearest recompression centre.
- Unlimited buddy requests, choice of male/female when available.
- Membership card accepted worldwide.
- Free copy of quarterly publication.
- Free sticker for your cylinder.
- Free video rentals on worldwide dive destinations.
- Members are invited to dive with dive clubs worldwide.
- Discounts at resorts/scuba centres worldwide.

DEI also collects valuable information on scuba centres, yacht charters, villa rentals, and dive resorts around the world. DEI serves as a clearing house of dive travel information with a keen eye on quality and safety. For further information contact: Divers Exchange International, PO Box 2382, Tisbury, MA 02568-9998, USA; phone: (617) 723 7134, fax: (617) 227 8145.

**diver's flag.** See '**Diver Below**' **flag**.

**dive slate.** A small flat piece of light coloured plastic with attached pencil, used for note keeping or writing messages to your buddy whilst underwater.

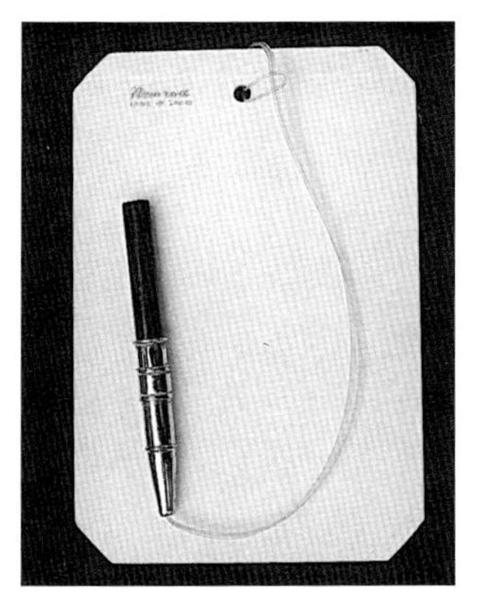

**Dive slates are handy for photographers to keep note of their camera settings.**

**dive tables.** See **DCIEM Sport Diving Tables**, **RNPL**, **The Wheel Dive Planner** and **US Navy Standard Air Decompression Tables**.

**dive watch.** An electronic or mechanical timing device which is sealed to prevent the entry of water. The dive watch is essential equipment for scuba divers, enabling bottom times and decompression times to be monitored accurately. The latest in dive watches has a depth gauge incorporated into the design. Some of the latest dive gauge combinations have a bottom timer and watch included.

Dive watches are usually larger and heavier than ordinary watches.

**diving accidents.** See **Diving Emergency Service** and **Project Stickybeak**.

**diving clubs.** The first amateur diving club in the world—The Bottom Scratchers—was formed by Glen Orr in 1933 in California, USA.

**Diving Emergency Service (DES).** A service available 24 hours a day to divers throughout Australia and overseas. In Australia ring toll free **(008) 088 200** or overseas **61 8 223 2855**. Your call will be taken by a specialist diving doctor who will ask a number of questions such as: How many patients? Is patient conscious and breathing? Diving profile? What is the location? The DES doctor will offer immediate first aid advice if necessary, and will contact the nearest hyperbaric facility to you and advise it of the situation and your contact number. The facility should be the next people to make contact with you and they will organise a retrieval if necessary. The Diving Emergency Service (DES) is administered by the Royal Adelaide Hospital. It began in 1984 as an initiative of the Australian Underwater Federation and the Department of Defence (Navy). The service is provided by Australian doctors and members of the defence forces who are expert in diving medicine. The Royal Adelaide Hospital Hyperbaric Medicine Unit provides the telephone facilities for the toll-free phone number. During 1986/87 the toll-free phone number registered 140 calls and the number of calls has increased dramatically to over 1000/year currently. The perceived need for the DES arises primarily from the very limited medical expertise in diving medicine throughout Australia. This has resulted in diving related conditions not being recognised and in divers dying or being left with long term handicaps unnecessarily.

Some of the running costs of the DES are covered by contributions from the major diving organisations, NASDS, NAUI and PADI.

In cases of diving accidents in the countries listed below (no connection with DES) you can call the following telephone numbers:

• New Zealand: Royal New Zealand Naval Hospital, Auckland, phone: (09) 454 000.

• The United States: Diver Alert Network (DAN), phone: (919) 684 811.

• The United Kingdom: HMS Vernon, Portsmouth, phone: (0705) 22351, ext: 2375/7, or ext: 2413/4/5 after hours.

**diving medical examination.** Examination of a person intending to learn to dive by a doctor qualified in this field, having at least the Diploma in Diving and Hyperbaric Medicine awarded by the South Pacific Underwater Medical Society. The normal family doctor is not trained or equipped to conduct the necessary examination that tests a person's fitness to dive. The diving medical examination should be carried out according to Australian Standard AS2299–1979. A diving medical examination form meeting these standards is now available from NASDS, Unit 7, 15 Walters Drive, Osborne Park, WA 6017; phone: (09) 244 3500; fax: (09) 244 1462. A routine chest X-ray should form part of this examination. On the successful completion of this examination a 'Certificate of Fitness' form can then be issued. It should be noted that there are at least three medical conditions which most certainly will exclude a person from scuba diving, namely asthma, diabetes and epilepsy. See **asthma**, **diabetes** and **epilepsy**.

**diving officer.** The person in charge of diving activities in a club or on a boat.

**diving platform.** A flat wooden or metal platform normally placed at sea-level at the stern of a boat enabling divers to enter and exit the water easily. Also called a duck-board. In professional operations any fixed or anchored amenity which diving is carried out from.

**diving reflex.** An automatic lowering of the heart-rate when a person's face is submerged in water. This natural phenomenon occurs in humans and in other mammals.

**dolphin.** A small toothed whale and member of the order Cetacea. The most familiar dolphin to Australians is the bottlenose dolphin *Tursiops truncatus* which is the species normally captured for oceanariums. Dolphins can swim at up to 50 km per hour and can dive to depths of about 100 metres. Approximately 13 species are known to live in Australian waters including: the common dolphin *Delphinus delphius,* the dusky dolphin *Lagenorhynchus obscurus,* the grey dolphin *Grampus griseus,* the humpback dolphin *Sousa chinensis,* the shortsnout dolphin *Lagenodelphis hosei,* the snubfin dolphin *Orcaella brevirostris* and the striped dolphin *Stenella coeruleoalba.* See **Monkey Mia**.

**Dolphin Information Centre.** See **Monkey Mia**.

**dolphin kick.** A leg kick using both legs together in one harmonious movement. A useful method of swimming if you happen to lose one fin, as this method does not unbalance the diver.

The bottlenose dolphin *Tursiops truncatus* is the species normally captured for oceanariums.

**dome port.** A dome shaped glass or acrylic housing used on the front of an underwater camera housing. Distortion of the photographic image can occur when using wide angle lenses with a standard flat port—using a dome port eliminates this distortion and increases the angle of coverage of the lens being used.

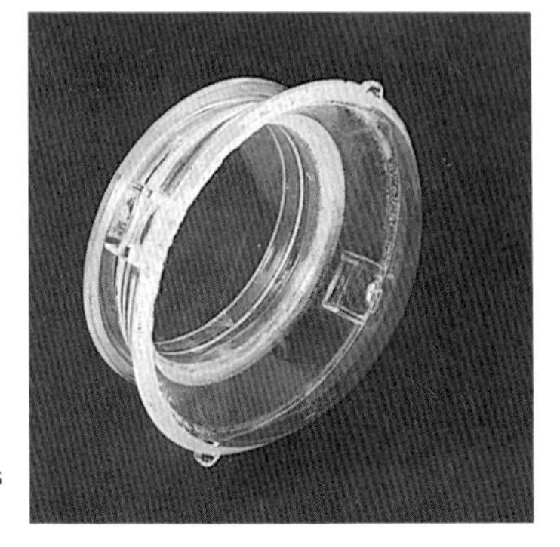

Dome ports for camera housings allow wide angle lenses to be used underwater with little or no vignetting.

**dorsal.** Situated at or near to the back of the animal i.e. the side of the animal which is normally directed upwards. Opposite of ventral. See **ventral**.

**DOT.** Department of Transportation. The DOT in the United States of America, is the public authority responsible for the laws and regulations governing the manufacture and use of scuba cylinders.

**downstream valve.** A valve in the second stage (mouthpiece) of a regulator which is held shut by a powerful spring and which holds back the compressed air in the hose coming from the first stage. A rubber diaphragm via a system of levers moves this valve, allowing air to flow into the regulator mouthpiece and hence into the lungs of the diver. Also called lever valve. See **upstream valve**.

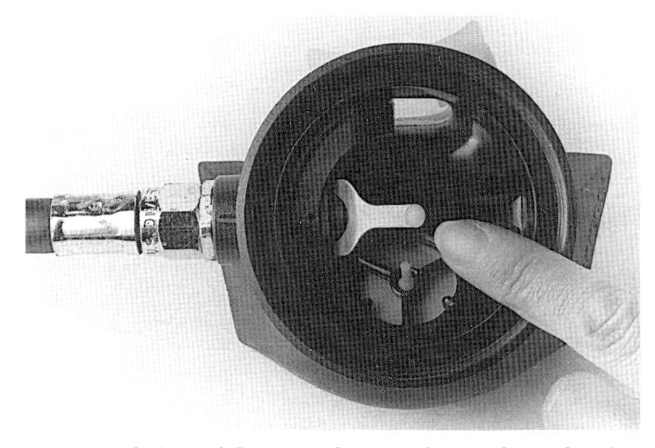

An internal view of the second stage of a regulator showing the lever or downstream valve.

**draft.** The vertical distance between the water surface and the lowest point on the keel of a boat.

**drift diving.** Diving in a strong current in the direction of the current flow. All divers should kit-up and be ready to enter the water together. The dive operator drops off all the divers at the same location and then follows the divers' exhaust bubbles and/or a surface buoy towed by the dive leader and collects the divers downcurrent on completion of the dive. Drift diving can be hazardous, so a strict dive plan is essential, as is a good skipper to tender the boat. It is essential with this type of diving that all divers remain close together.

Divers' names should be checked off a list on entering the water and at the completion of the dive to avoid the possibility of leaving anyone behind.

**The white water in this photograph was caused by reversing the boat's engines to allow divers drifting with the current to catch up with the boat quickly and enable them to leave the water safely.**

**drift netting.** A technique of fishing where kilometres of nets are hung vertically in the water suspended by floats and left free to drift about until retrieved. This technique is non-discriminatory, catching everything from seabirds to whales and sharks. Often referred to by conservationists as 'walls of death'.

**drugs from the sea.** Biologically active chemicals produced by marine organisms and exploited by scientists to treat human and animal diseases. A search is being carried out worldwide for new drugs from the sea. Such chemicals have the potential to be useful in human health, agriculture, industry and veterinary applications. The Australian Institute of Marine Science situated in Townsville, Queensland, along with various universities and research organisations, is actively involved in this research.

**dry suit.** An underwater exposure suit used in cold water, in which the diver remains dry. This type of suit is made of thin sheet rubber–composite material and is worn over insulating undergarments. Air is pumped into the suit from a low pressure hose attached to the divers' regulator in order to help control descent and ascent and to prevent chaffing of the skin from suit squeeze. A weight belt and ankle weights are normally used with this type of suit. Special training is required to use dry suits safely. See **ankle weight**.

Scientists at work collecting and identifying marine organisms which are a potential pharmacopoeia of new and useful chemical substances which may help fight diseases as diverse as malaria and AIDS.

**duck-board.** See **diving platform**.

**duck dive.** A free-swimming surface dive achieved by bending the waist and extending the legs to a vertical position. The weight of the extended legs causes submergence.

**dugong.** An aquatic mammal and member of the order Sirenia, family Dugongidae. The dugong *(Dugong dugong)* is an aquatic herbivorous mammal which lives in the warm shallow coastal waters of the Indo-Pacific region. Dugongs graze on pastures of sea-grasses and grow to three metres in length and about 400 kg in weight. Western Australia has the largest populations of dugongs in the world and a recent survey estimated the Shark Bay population to be in the region of 10000 animals. In the Northern Territory however, dugong numbers have been reduced dramatically due to traditional harvesting by aborigines and by accidental deaths from drowning when dugongs are caught in fishing nets such as those used by barramundi fishermen and shark meshing contractors. The dugong is one of only four surviving species of sirenians, or sea cows. The other species are known as manatees and are found in the Amazon River, the Caribbean region and in West Africa. Manatees are dependent on freshwater environments of rivers and estuaries, whereas the dugong is strictly marine.

**dugong-grass.** A common name (used mainly in Queensland) for a range of sea-grasses grazed by dugongs. See **sea-grass**.

**duration of dive.** See **bottom time**.

**dysbaric osteonecrosis.** Localised death of bone tissue in the hips, knees and shoulders of scuba divers, thought to be caused by tiny bubbles of nitrogen. Sometimes called a number of common names including; bone necrosis, bone rot and aseptic bone necrosis. This is a disease brought about by diving. Although the exact cause of the disorder is still uncertain it is thought to be a long term manifestation of decompression sickness. The main effect of the disease is the localised death of the bone in joints, usually the hips, knees or shoulders. In bad cases, it eventually leads to the total break down of the joint/s. This disease can be detected by x-ray, or the more modern and safer alternatives of magnetic resonance imaging and technetium bone scans. To avoid bone necrosis, avoid repeated deep diving and use your dive tables conservatively.

***Dysidea herbacea.*** A small marine sponge and member of the class Demospongiae, order Dictyoceratida. This sponge is commonly found in the tropical Indo-Pacific Ocean and is usually grey-green in colour, encrusting in habit, with small erect 'ears' which may have yellowish tips. Its tissues contain numerous endosymbiotic cyanobacteria *(Oscillatoria symbiotica)*. See **sponge**.

**The small grey-green sponge *Dysidea herbacea* is widespread in the Indo-Pacific Ocean.**

**EAR.** Acronym for Expired Air Resuscitation. See **artificial respiration**.

**ear clearing.** The act of equalising air pressure in the middle ear with the pressure of the surrounding water. A number of techniques can be employed to achieve this including: swallowing, yawning, moving the lower jaw, and the Valsalva manoeuvre (pinching the nostrils and blowing gently). See **ear plugs**.

**ear plugs.** Rubber devices used by swimmers to prevent water entry to the ears. These are extremely dangerous if used underwater by snorkellers or scuba divers. The surrounding water pressure will drive the ear plugs deep into the ear spaces and can cause pain, bleeding and ruptured eardrums.

**Earthwatch Australia.** A non-profit research organisation which coordinates volunteer teams of workers to help scientists and teachers involved in field research projects in various countries. Volunteer certified scuba divers are needed to help with much of this research. The following information has been supplied by Earthwatch Australia:

Earthwatch is not a holiday, it is hard work, and you pay to do it. An unusual concept but one that works well: paying volunteers underwriting the costs of field research while donating their time, intelligence and labour for two weeks on a scientific research expedition.

It is a company of scholars and citizens working together to preserve the world's endangered species and habitats, to explore the oceans and heritage of its peoples, to promote international cooperation and to search for solutions concerning a changing planet. Everyone benefits—the scientist receives funding and labour; the volunteer, a chance to learn, discover, to contribute and to do something really different.

In any one year there are over 100 expeditions in about 44 countries. The volunteers pay a 'share of costs' (anything from $800–$2000) which covers their food, accommodation, and use of equipment in the field, during the two week expedition.

For the marine minded, Earthwatch has a selection of exciting projects; in Bonaire, (Caribbean) certified scuba divers map the coral reefs, diving twice a day on a tropical island. Snorkelling volunteers work off Belize following squirrelfish and ultrasonically 'tagged' moray eels, or work in the freshwater caverns—known as blue holes, on islands in the Bahamas. Studies of the Red Sea reef fish are undertaken regularly from Eliat, on the Israeli coast, and a short while ago scuba diving volunteers from Australia and overseas worked together on the incredibly beautiful Lizard Island in the Great Barrier Reef, trying to find what triggers the largest female damsel fish to turn herself into a male.

As well as snorkelling and diving projects, Earthwatch also funds a range of marine mammal studies. Studying whales takes exceptional commitment, especially when you consider that a female orca can live to more than 100 years. Working from boats, volunteers photograph and document the behaviour of orcas (killer whales) in Puget Sound, (Washington State USA). Scientists are using results collected from this particular study to address questions raised after the *Valdez* oil spill, such as the effects of pollution, and development, on orca populations.

In the Hawaiian islands, the songs of the humpback whale are being recorded. Some of the melodies last for hours and cover a range that surpasses human hearing. With the help of enthusiastic volunteers, scientists are trying to find out why these huge ocean creatures sing, and what the different songs mean. On the Mexican coast, humpback whale breeding grounds are threatened from the development of multi-million dollar complexes. To be able to protect this species more information is urgently needed on their requirements and lifestyles. Volunteers are needed to go to Mexico and assist with collecting this crucial information.

The wild dolphin populations around Sarasota Bay in Florida need further studies to understand their complex social behaviour. For 18 years 100 dolphins have been tracked and family relationships are now becoming clear. This year, studies will be starting on new family groups of wild dolphins still located in Florida, but at a different bay.

Further south, on the Pacific island of Tonga, snorkelling volunteers have been mapping the distribution of the giant clams—a vital part of the local people's diet and folklore. They have also been helping to promote the marine conservation ethic by working closely with the Tongans.

There are many more projects associated with the sea, including some unlikely ones; studying the rise and fall of the Caspian Sea while working from a Soviet research vessel, or participating in national coastal surveys along Britain's 8000 km coastline. The opportunities are there, why not take them. Earthwatch Australia, 1st floor 457 Elizabeth Street, Melbourne, Vic. 3000; phone: (03) 600 9100, fax: (03) 600 9066.

**ebb tide.** Low water, caused by the flow of water away from the shore.

**echinoderm.** Member of the phylum Echinodermata, a group of marine animals comprising the sea-urchins, sea-stars, brittle-stars, feather-stars, and sea-cucumbers. Echinoderms typically have a five rayed symmetry, a calcareous skeleton which is found in the connective tissue layer of the skin, and a water vascular system that moves their tube feet by changes in hydrostatic pressure. Most echinoderms are harmless to touch but several species of sea-urchins have poisonous spines and/or pedicellariae which can cause painful stings. See **feather-star**, **sea-cucumber**, **sea-star**, **sea-urchin** and **pedicellariae**.

**echo sounder.** See **depth sounder**.

**ecology.** The study of living organisms in their environment.

**ectoparasite.** An organism which lives parasitically on the outside of another organism.

**eddy.** A swirling or rotary current of water.

**Edgerton, Harold. E.** An inventor who made the first electronic flashguns for photography. In 1956 he invented an underwater robotic camera and electronic flash. This equipment was used to photograph features on the bottom of the Romanche Trench in the Atlantic Ocean at a depth of 7620 metres.

**Edwards, Milne. Professor.** Supposedly the first biologist to use a diving helmet, in 1844. He was supplied by air pumped from the surface, to several metres beneath the waters of the Mediterranean Sea.

**eel-grass.** See **Zostera**.

**electric ray.** See **numbfish**.

**elephant snail.** Mollusc and member of the family Fissurellidae (false limpets). The elephant snail is one of the most easily recognised molluscs of the Australian seashore. Two species are found along rocky coastlines in Australia: *Scutus antipodes* and *Scutus granulatis.* Both are nocturnal algal browsers and are found in crevices and in rock pools from the low tide mark down to about 20 metres. They usually grow to about eight centimetres but may reach 15 cm in length and are found from Queensland to Tasmania. In the past they were a favourite food of the coastal Aborigines.

**emergency radio beacon.** See **EPIRB**.

**emergency services for divers.** See **Diving Emergency Service**.

**emphysema.** The abnormal presence of air or gases in the body tissues.

**entering the water.** Shore divers can simply walk into the water from a sandy beach or use a giant stride entry technique with the feet wide apart if entering the water from an elevated position. Boat divers can use a number of different methods of entering the water depending on the type of boat they are diving from—giant stride, forward roll, back roll, front jump (feet together). See **exiting the water**.

**epaulette shark.** Member of the order Orectolobiformes, family Hemiscylliidae. The epaulette shark, *Hemiscyllium ocellatum,* is a small shark (harmless to humans) which grows to one metre in length and is abundant on the coral reef flats of Australia's tropical north and the waters of the Indo-Pacific. Its common name is derived from the two large white edged black spots resembling epaulettes just behind the pectoral

**The epaulette shark *Hemiscyllium ocellatum* is a small harmless species which is often left stranded by the tide in coral pools.**

fins. During the day the epaulette shark remains hidden under coral boulders, but it comes out of hiding at night to search the reef flats for worms, shrimps, crabs and other small invertebrates.

**epilepsy.** A medical condition characterised by episodes of loss of consciousness and convulsive seizures due to uncontrolled electrical discharges from the nerve cells in the brain. People who suffer from epilepsy should never scuba dive.

**EPIRB.** Emergency Position Indicating Radio Beacon. Small, waterproof, buoyant emergency location transmitters that transmit a signal on the two aeronautical distress frequencies about every two seconds until switched off or until the battery runs down. EPIRBs are not a substitute for a marine radio as the EPIRB frequencies are only monitored by certain types of aircraft. See **boating safety equipment**.

**An Emergency Position Indicating Radio Beacon or EPIRB.**

**equalising pressure.** As a diver descends, water pressure increases. This causes no problems for most parts of the human body except where their are air spaces. If left unequalised, pressure can be felt building in the ears, resulting eventually in a burst ear drum. Unequalised pressure has the effect of pushing the dive mask against the face, which can cause the eyes to bulge resulting in the rupture of small blood vessels (see mask squeeze). The easiest way to equalise pressure in the ears is to grip the nose with the thumb and forefinger and blow gently (Valsalva manoeuvre). Some people find it easier to swallow which can also achieve the same result. The problem with the face mask can be simply resolved by blowing air gently through your nose into the mask. See **mask squeeze**.

**estuary.** An inlet on the coastline where a river outflow meets the sea.

**exhale.** To breathe out.

**exiting the water.** Leaving the water from the surf or rocks can be hazardous. The following methods are suggested:

• Surf beach exits. After making a surface assessment of the wave pattern swim underwater as close as possible to the edge of the breaking waves then wait for a breaking wave and follow in behind it. Don't loiter in the shallows, get out of the water as quickly as possible.

• Rocky shoreline exits. Getting out of the water onto rocks through turbulent surf should be avoided if at all possible. It's a good idea to have a person waiting on the rocks to help you leave the water and call for assistance if required. If there is no other way, wait for a lull in the wave pattern then follow a small wave in just behind the breaking crest, before it begins to subside get a firm grip on the rocks and brace yourself against the backsurge (the water running back to the sea). As soon as the backsurge weakens, climb quickly for higher ground before you are pounded by the next wave. If your dive plan requires an exit onto rocks, take a length of rope and an extra lead weight with you. The rope can be lashed to a rock and the lead weight used to hold the other end of the rope in the water so you can utilise it for your exit. One of the most important rules of diving to remember is to dive within your own personal limitations and never exceed them.

**expiration.** The act of breathing out air from the lungs.

**face mask.** A mask worn on the face so that the diver's eyes and nose are in air rather than water, allowing a clear view underwater. Cressi of Italy invented the forerunner to the modern face mask; it was called the 'Pinocchio'. It provided an easy way of equalising pressure by squeezing the nose with the thumb and forefinger. Facial contours vary greatly among individuals and races and one type of mask will not be suitable for everyone. The easiest test to apply in order to see whether or not a mask is suitable for your particular face is as follows: without using the strap, place the mask against your face and gently breathe in through your nose; if the mask remains in place it will be suitable for you. If you have a big nose and/or a full beard you will find it difficult to obtain a face mask that fits perfectly. Black rubber face masks are rapidly being replaced with clear silicone rubber masks which are softer, have less volume, are resistant to ultraviolet light and look better in photographs. Full face masks are also available but have to be used with special regulators and are very expensive. They are used mainly by professional divers who need to communicate with the surface dive tender. If you wear spectacles, prescription lenses can be fitted by your optometrist (OPSM specialises in fitting these lenses), and a range of standard optically-corrected lenses is available from some dive shops. One problem associated with clear silicon rubber face masks and underwater photographers is reflections inside the mask making it difficult to see clearly into camera viewfinders: so if you take your underwater photography seriously, a black or grey opaque rubber mask is a better choice.

**A selection of face masks should be tried for fit and comfort as individual faces vary in their shape, size and profile.**

**fathom.** A measurement of water depth equal to 1.83 metres or six feet.

***Fathom* magazine.** A quality colour magazine dedicated to scuba divers and spearfishermen, first published by Gareth Powell Associates Pty. Ltd. in Sydney in 1970. Ten issues were printed over a three year period. The editor was John H. Harding, a well known Australian underwater cinematographer famous for his 'Underwater Seafari Films' which were narrated live. The magazine is now considered a collector's item.

*Fathom* magazine was one of the first colour dive magazines published in Australia.

**FAUI.** Federation of Australian Underwater Instructors. See **National Association of Scuba Diving Schools Australasia Inc. (NASDS)**.

**feather-star.** Echinoderms in the class Crinoidea in which the body is supported on a ring of jointed appendages called cirri which originate from the underside of the body cup. Most feather-stars have five main arms which may branch into as many as 200 sub-branches each of these has jointed finger-like appendages called pinnules fringing the two sides of each arm. When extended, this array of arms and pinnules acts like a fishing net trapping plankton which is transferred via ciliated grooves to the mouth. The feather-star *Comanthus trichoptera,* is probably the most wide-ranging species in Australian coastal waters and occurs in a variety of colours including: orange, green, brown or black. See **echinoderm**.

**Feather-stars are often seen perched on gorgonian corals, where they are exposed to the passing tidal flow that carries plankton on which these animals feed.**

**Federation of Australian Underwater Instructors.** See **National Association of Scuba Diving Schools Australasia Inc. (NASDS)**.

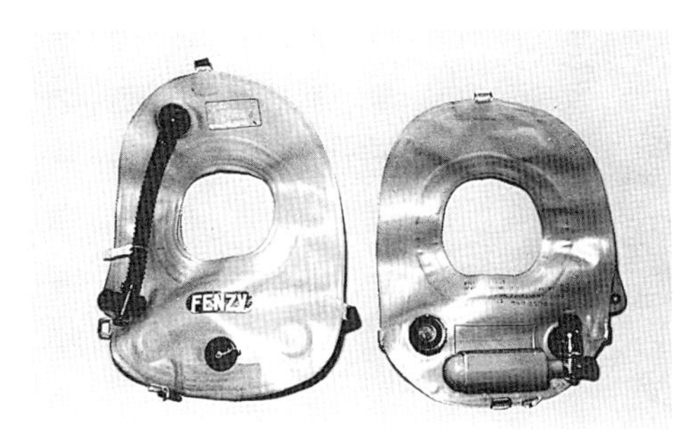


**The old style horse collar design buoyancy compensators such as the Fenzy model have been replaced with more modern designs.**

**Fenzy.** A brand of buoyancy control device (BCD) which incorporates a small cylinder of compressed air for emergency use. The fenzy is of a horse collar design and fits over the diver's head, being secured by straps around the waist and between the legs. Buoyancy control is achieved by blowing into a mouthpiece connected via a rubber hose to the vest. Air is released by holding the hose over the head and depressing a button connected to the mouthpiece. Nowadays these units have been replaced with scuba-fed jackets or vests that incorporate a back pack and fit directly onto a scuba cylinder. See **buoyancy compensator**.

**fins.** Swimming fins (also called flippers) worn on the feet to aid swimming, invented in 1929 by Frenchmen Louis de Corlieu. Fins are available in many shapes, styles and colours but there are two basic types; the shoe type and the open heel type. The shoe type fin requires no maintenance other than a wash in fresh water. A thin pair of soft soled rubber booties or a pair of sox can be worn with these to stop chaffing. The open heel type is the most popular with scuba divers but has straps and clips that require adjustment and replacement at intervals. Hard soled booties are normally worn with this type of fin. See **booties**.

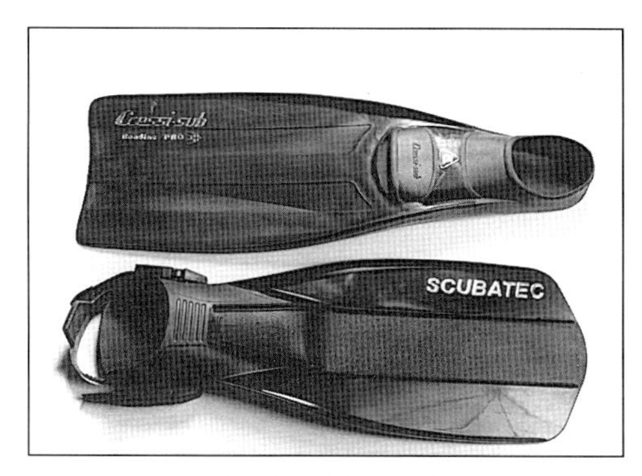

**Open heel and shoe style fins.**

**fire-coral.** Non-scleractinian coral and member of the class Hydrozoa, family Milleporidae. Fire-coral can cause a burning or stinging sensation on the softer parts of human skin. See **hydrocoral** and **hard coral**.

**firefish.** See **zebra lionfish**.

**first aid.** The initial care of the injured or the sick. Detailed first aid is beyond the scope of this publication but the following references are recommended:
• *Australian First Aid. The Authorized Manual of St. John Ambulance Australia;* Ruskin Press North Melbourne.
• *Diving and Subaquatic Medicine* by Carl Edmonds, Christopher Lowry and John Pennefather; The Diving Medical Centre, Sydney 1976.
• *Marine Animal Injuries to Man* by Dr Carl Edmonds; Wedneil Publications 1984.
• *The DES Emergency Handbook* by John Lippmann and Stan Bugg; J. L. Publications, PO Box 381,Carnegie, Vic. 3163.

**first stage.** Part of a regulator which attaches to a scuba cylinder (containing high pressure air) and reduces the cylinder pressure to a line pressure of about 827 kPa (120 psi) which is fed via a hose into the second stage of the regulator (part with a mouthpiece). See **regulator**.

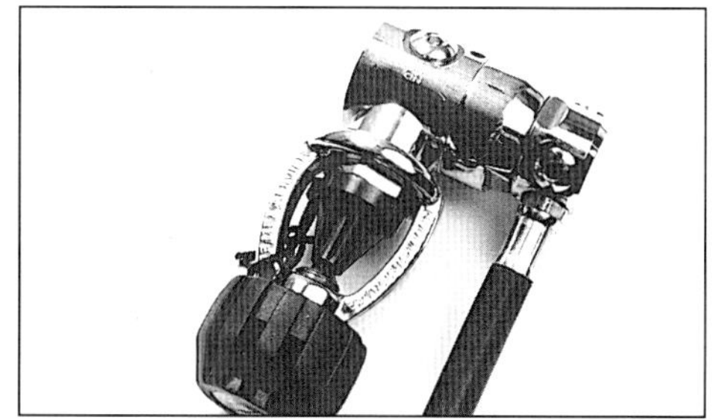
**The first stage of a scuba regulator.**

**first underwater photograph.** The first underwater photograph was taken by two Englishmen, Mr William Thompson and Mr Kenyon in February 1856.

**fish.** Cold-blooded aquatic vertebrates, breathing with gills and swimming with the aid of fins. A collective term including: jawless fish, class Agnatha; cartilaginous fish, class Chondrichthyes and bony fish, class Osteichthyes.

**fish books.** Books which contain information useful for identifying fishes, their habits and life cycles. The following is a list of useful reference material, though it is by no means complete:
- *Australian Fisherman's Fish Guide* by Neville Coleman; Bay Books, Sydney 1978.
- *Australian Sea Fishes North of 30 degrees South* by Neville Coleman; Doubleday, Sydney 1981.
- *Australian Sea Fishes South of 30 degrees South* by Neville Coleman; Doubleday, Sydney 1980.
- *Butterfly and Angelfishes of the World* Volume 1 by Roger C. Steene; A.H. and A.W. Reed, Sydney 1978.
- *Checklist of Fishes* by Barry C. Russell; The Great Barrier Reef Marine Park Authority, Townsville, Queensland 1983.
- *Dangerous Fishes of Western Australia* by J.B. Hutchins; Western Australian Museum 1980.
- *Fishes of the New Zealand Region* by Wade Doak; Hodder and Stoughton, Auckland New Zealand 1972.
- *Fishes of Western Australia* by Dr Gerald R. Allen; T.F.H. Publications 1985.
- *Grant's Fishes of Australia* by E.M. Grant; E.M. Grant Pty Limited, Queensland 1987.
- *Guide to Fishes* by E.M. Grant; (5th ed.) Dept. Harbours and Marine, Brisbane 1982.
- *Handbook of Australian Fishes* by Ian S.R. Munro; Reprinted from *Fisheries Newsletter* 1960-1961, by Publicity Press Ltd., Sydney.
- *Hawaiian Reef Animals* by Edmund Hobson and E.H. Chave; University Press of Hawaii 1973.
- *Marine Fishes of Australian Waters* by Kenneth Pulley; Lansdowne Press 1974.
- *Rainbowfishes of Australia and Papua New Guinea* by Dr Gerald R. Allen and Norbert J. Cross; Angus and Robertson, Australia 1982.
- *Reproduction in Reef Fishes* by Dr. R.E. Thresher; T.F.H. Publications, Hong Kong 1984.
- *Sea Fishes of Southern Australia* by Barry Hutchins and Roger Swainston; Swainston Publishing, Perth 1986.
- *Shadows in the Sea* by Harold W. McCormick, Tom Allen and Captain William Young; Weathervane Books, New York 1963.
- *Sharks* consulting editor John D. Stephens; Golden Press, Sydney 1987.
- *Sharks* by Peter Goadby; Ure Smith, Sydney 1975.
- *Sharks of Australia* by G.P. Whitley; Jack Pollard Publishing, Sydney 1981.
- *Sharks Silent Hunters of the Deep;* Reader's Digest Services, Surry Hills, NSW 1986.
- *Skin Diver Magazine's Book of Fishes* by Hillary Hauser; Pisces Book Co., Inc. 34 West 32 Street New York, NY 10001, USA 1984.
- *The Australian Fisherman's Companion* by Harold Vaughan; Lansdowne Press, Sydney 1986.

• *The Complete Diver's and Fishermen's Guide to Fishes of the Great Barrier Reef and Coral Sea* by John E. Randall, Gerry R. Allen, Roger C. Steene; Crawford House Press, Bathurst NSW 1991.
• *The Fresh & Salt Water Fishes of the World* by Edward C. Migdalski; Greenwich House, New York 1989.
• *The Fishes of New Guinea* by Ian S.R. Munro; Department of Agriculture, Stock and Fisheries, Port Moresby, Papua New Guinea 1967.
• *The Fishes of Rottnest Island* by Barry Hutchins; Creative Research, Perth 1979.
• *The Marine and Estuarine Fishes of South-Western Australia* by Barry Hutchins and Martin Thompson; Western Australian Museum 1983.
• *The Marine Fishes of North-Western Australia* by Gerald R. Allen and Roger Swainston; Western Australian Museum 1988.
• *Trawled Fishes of Southern Indonesia and North Western Australia* by Thomas Gloerfelt-Tarp and Patricia J. Kailoa; The Australian Development Assistance Bureau.
• *Tropical Reef-Fishes of the Western Pacific: Indonesia and Adjacent Waters* by Rudie H. Kuiter; Aquatic Photographics PO Box 124, Seaford, Vic. 3198.
• *Watching Fishes* by Robert Wilson and James Q. Wilson; Harper & Row, New York 1985.

**fish identification posters.** Posters depicting fishes, useful for decoration and identification purposes. Wall posters are available from the following sources:
**1. Diver Publications**, 9 Ross St, Mitcham, Vic. 3132. This company sells waterproof marine identification posters featuring Australian fishes. Posters are available from selected dive shops. The titles are as follows:
• *Common Marine Fishes of Southern Australia,* featuring 60 fishes.
• *Common Marine Creatures of South-Eastern Australia,* featuring 48 marine animals.
**2. First Assistant Secretary**, Department of Fisheries and Marine Resources, PO Box 417, Konedobu, Papua New Guinea. The name and description of the poster follows:
• *Fishes of Papua New Guinea and Northern Australia.* The poster features 72 colour prints of fishes in family groupings, each fish is labelled with its scientific name and maximum size.
**3. The New Zealand Underwater Association** produces identification posters and these are available from; PO Box 875 Auckland 1 New Zealand; phone: 0011 649 849 5896. The titles are as follows:
• *New Zealand Fishes.*
• *New Zealand Deep Water Fishes.*

**fish-lice.** Crustaceans and members of the subclass Branchiura. Fish-lice are free-swimming parasitic crustaceans and are related to copepods and barnacles; they have flattened bodies and no gills. These ectoparasites use their hooked feet to cling to the skin and gill cavities of fishes and their sucking mouth parts are used to feed on the blood and mucus of their hosts. See **ectoparasite**.

**flare.** See **distress flare**.

**flippers.** See **fins**.

**float line.** A long piece of floating rope attached to a small buoy. The whole device trails from the stern of a moored boat in order to assist divers to reach the boat when a current is running. A float line can also be thrown to divers in distress. It is also known as a mermaid line.

**flood tide.** The tide at its highest level.

**flotsam.** A ship's cargo or wreckage found floating on the ocean. See **jetsam** and **ligan**.

**focsle.** The crew's quarters in the foreward part of a ship.

**following sea.** A sea that overtakes a boat possibly causing it to broach (turn broadside to the wind and waves).

**free ascent.** Swimming to the surface without the use of scuba. A free airway must be maintained to allow expanding air to escape from the lungs. This is a very dangerous procedure and should only be attempted in an emergency.

**fresh breeze.** A wind of 17–21 knots or 31–39 km/hr.

**Friends of Marine Parks.** A voluntary support group for marine parks and reserves in Victoria. For membership Information, contact: Mr Colin Cook, 27 Raleigh St, Thornbury, Vic. 3071.

**fringing reef.** Coral reef attached to the mainland or a continental island. The majority of the resort islands on the Great Barrier Reef are located on fringing reefs. Lord Howe Island has a fringing reef which contains the southernmost coral reef in the Pacific region. Fringing reefs also occur close to the mainland along the Daintree coast (Queensland), and North West Cape (Western Australia). See **Ningaloo Marine Park**.

**frogman.** Any diver wearing scuba gear (including mask and fins). The term was coined by newspapermen during the Second World War to describe navy divers who used oxygen rebreathers. See **closed circuit breathing apparatus**.

**fugu.** An expensive seafood delicacy of beautifully arranged, thinly sliced pieces of pufferfish. In Japan chefs are specially trained to eliminate most of the toxic parts of the pufferfish and are licensed to prepare the flesh for human consumption; occasional mistakes result in severe illness or even death. Fugu is an expensive but popular dish with Japanese gourmets who pay approximately 20000 yen ($A220) per plate. Up to 2000 tonnes of pufferfish are consumed each year in Japan. Scientists are worried about depletion of fish stocks and in order to overcome this problem they have begun artificially hatching eggs and releasing fish to restock the sea. Attempts are also being made to raise puffer-fish in fish farms to relieve the pressure on wild stocks. See **pufferfish**.

**Gagnan, Emile.** Co-inventor of the scuba demand valve with Jacques-Yves Cousteau. The first prototype was produced in Paris in 1943. Gagnan adapted a regulator originally designed for gas-powered cars.

**gale.** Winds of 34–40 knots or 63–74 km/hr.

**gas laws.** See **Boyle's**, **Charles'** and **Dalton's Laws**.

**gastropod mollusc.** Mollusc and member of the class Gastropoda, examples are abalone, limpets and sea-slugs.

**gauge pressure.** Indicates the internal pressure in a closed vessel such as a scuba cylinder. See **contents gauge**.

**gentle breeze.** A wind of 7–10 knots or 13–18 km/hr.

**giant clam.** See **clam**.

**giant coxcomb-oyster.** Bivalve mollusc and member of the family Gryphaeidae. The giant coxcomb-oyster *Pycnodonte hyotis,* is the largest edible oyster in Australia, growing to about 20 cm in diameter. It is found in tropical waters from the low tide mark down to about 20 metres. See **mud-oyster**, **Sydney rock-oyster**, **Pacific oyster** and **midden**.

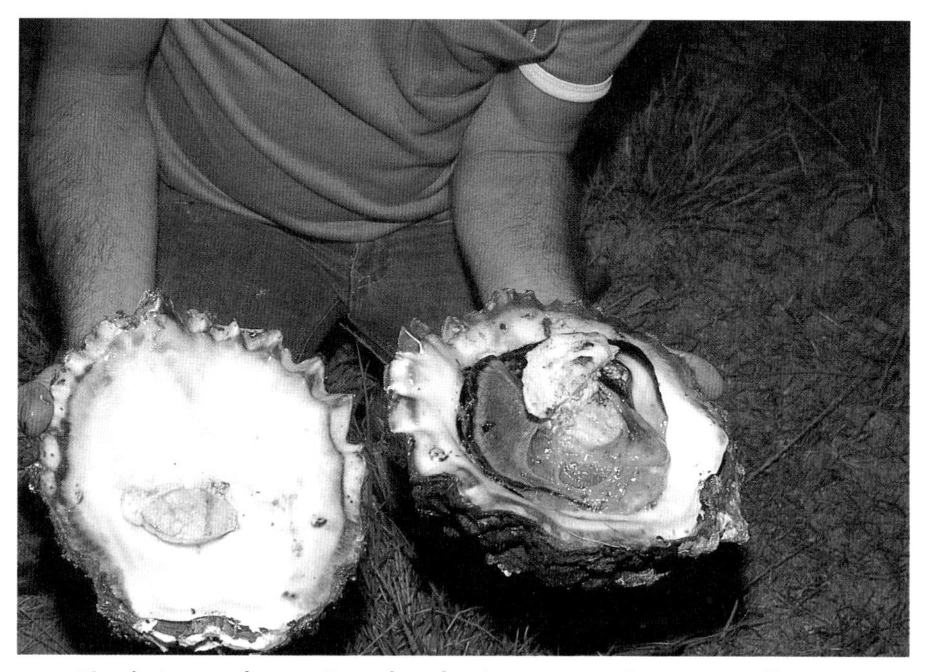

**The giant coxcomb-oyster *Pycnodonte hyotis* can grow to about 20 cm in diameter.**

**gidgee.** An Aboriginal spear. This term is used by spearfishermen when referring to a handspear or Hawaiian sling. See **handspear** and **Hawaiian sling**.

***Glaucus atlanticus.*** A pelagic nudibranch suborder Aeolidacea, family Glaucidae. *Glaucus atlanticus* is found floating on the surface in all warm oceans of the world and is often washed ashore after consistent onshore winds along with other floating coelenterates such as *Porpita porpita, Velella velella* (by-the-wind-sailor) and *Physalia utriculus* (Portuguese man-of-war). *Glaucus atlanticus* grows to about five cm in length and feeds on the floating coelenterates (mentioned previously), incorporating their stinging cells into its own body tissue, where they are stored until required for self-defence. As a consequence this animal can cause mild irritation to the softer areas of human skin when touched. See **by-the-wind sailor**, ***Porpita porpita*** and **Portuguese man-of-war**.

**gloves.** Hand coverings used by divers to insulate against heat loss, and for protection against coral cuts, hydroid stings, sponges and other marine hazards. Gloves also have the additional advantage of giving a splash of colour to underwater photographs. Gloves are made from neoprene, synthetics, leather, cotton or a combination of these materials.

**Dive gloves are used for both tropical and cold water diving. They provide protection against abrasions, cold water and stinging organisms.**

**gobies.** Small bottom-dwelling fishes in the order Perciformes, family Gobiidae, found in warm and temperate seas. They are common inhabitants of marine rock-pools. Gobies have two separate dorsal fins. There are several freshwater species.

**goggles.** A term often incorrectly used for a modern underwater face mask. It is more correctly used for the small eye-cups worn by swimmers. Goggles should never be used for scuba diving because there is no way of equalising pressure and haemorrhaging of the blood vessels in both eyes could result. See **face mask**.

**Goggles can be used for surface swimming but never for scuba diving.**

**The pelagic nudibranch *Glaucus atlanticus* is unmistakable due to its bizarre shape and striking blue and silver colouration.**

**golf-ball sponge.** A marine sponge, class Demospongiae, order Hadromerida. Various species of the genus *Tethya* are commonly called golf-ball sponges because of their small rounded shapes. Colours range from orange and yellow through to brown, red and purple. They are attached to the substrate by small root-like appendages at their base. See **Porifera**.

The golf-ball sponge *Tethya* sp.

**goose barnacle.** Crustacean and member of the class Cirripedia, suborder Lepadomorpha. Goose barnacles, attached by short or long rubbery stalks to logs and other floating debris, are a common find for beachcombers. They foul anything floating in the ocean including: the bottom of ships, pylons, buoys and the underwater structure of oil rigs. Two common species of goose barnacles found in Australian waters are the long stalked *Lepas anatifera* and the short stalked *Lepas anserifera.* The common name is probably derived from the shape and length of the rubbery stalks which resemble the neck of a goose. See **barnacle**.

Goose barnacles (*Lepas* sp.) encrusting a mooring buoy.

**gorgonian.** Horny coral and member of the subclass Alcyonaria or Octocorallia, order Gorgonacea. Gorgonians are commonly called sea-fans or sea-whips: sea-fans are generally fan shaped while sea-whips consist of a single long branch. The main axis of a gorgonian has a central strengthening rod of gorgonin (a horny material) or is sometimes calcareous. Gorgonians are attached at their base to a hard substrate such as rock or coral. Gorgonian branches are often brightly coloured and are covered with retractable polyps. Some colonies may reach three metres in height and may have colourful feather-stars, hydroids, bryozoans, brittle stars or small cowrie shells adhering to the colony. See **soft coral** and **black coral**.

Gorgonian sea-fans may grow to three metres in height.

**Great Barrier Reef (GBR).** The largest system of coral reefs in the world, made up of nearly 3000 individual reefs and home to about 330 species of hard coral. The Great Barrier Reef stretches for more than 2000 km from Lady Elliott Island in the south to the tip of Cape York in north Queensland and beyond. Most of these coral reefs lie within the world's largest marine park which covers an area of 344000 $km^2$ which is an area larger than Victoria and Tasmania combined. This huge marine park is managed by the Great Barrier Reef Marine Park Authority in conjunction with the Queensland National Parks and Wildlife Service. On the 21 October 1981 the Great Barrier Reef was inscribed on the World Heritage list, in recognition of its outstanding universal value. The Great Barrier Reef complex hosts some of the best scuba diving sites in the world.

The Great Barrier Reef is made up of nearly 3000 individual coral reefs.

**Great Barrier Reef Marine Park Authority (GBRMPA).** A government funded organisation set up to suggest guidelines and advise the government on the protection and care of the Great Barrier Reef. The postal address is: PO Box 1379, Townsville, Qld 4810; phone: (077) 81 8811, fax: (077) 21 3445. If you require information about the GBR Marine Park or any other aspect of the reef, a whole range of publications is available from the aquarium shop attached to the Great Barrier Reef Wonderland Aquarium. A list of publications is available by writing to: The Manager Aquarium Shop, Great Barrier Reef Wonderland, Flinders Street East, Townsville, Qld 4810; phone: (077) 81 8875.

**Great Barrier Reef Wonderland Aquarium.** A complex housing 16 aquaria, conceived and operated by the Great Barrier Reef Marine Park Authority in Townsville, Queensland. The Great Barrier Reef Wonderland houses the world's largest coral reef aquarium, Australia's first Omnimax theatre, a branch of the Queensland Museum and a number of shops. The aquarium complex was a bicentennial commemorative project and was opened by the Prime Minister of Australia on the 24 June 1987. It provides an important means of educating the public about the Great Barrier Reef during the 1990s and beyond. The main tank simulates life on the Great Barrier Reef and has waves, tides, currents and a unique water purification system that relies on the photosynthetic action of marine algae to keep the water clean. This provides ideal conditions for the growth of corals. In fact the main tank contains a living coral reef, with many associated plants and animals. Recently the aquarium corals spawned at the same time as those in the outside reefs. A separate smaller tank displays reef predators including sharks and large fish. A special touch tank is popular with children and a audiovisual spectacular is featured in the Omnimax theatre, where a projected image is displayed on a domed screen 18.75 metres in diameter. The screen surrounds the audience of up to 200 people in a cone of light and sound and is an experience not to be missed. Great Barrier Reef Wonderland, Flinders Street East, Townsville, Qld 4810; phone: (077) 81 8875.

**green alga.** Member of the division Chlorophyta. Green algae vary in colour from yellow-green to almost black. There are many freshwater species as well as marine forms. Common marine genera, depending on the locality, include: *Caulerpa, Chlorodesmis* (turtle weed), *Codium, Enteromorpha* (baitweed) and *Ulva* (sea-lettuce). Green algae are found mainly in the upper intertidal zone on rocky shorelines especially in sheltered areas such as rock pools, harbours and estuaries and certain species seem to flourish in polluted waters and can be found where stormwater runs into the sea. See **caulerpa** and **turtle weed**.

***Codium fragile*** **is a common temperate water green alga.**

**Greenpeace.** An international organisation dedicated to preserving our earth and all the life it supports. The following information has been supplied by Greenpeace Australia:

Greenpeace was born from a sense of outrage at the way in which our fragile planet was, and still is, being abused by governments and industries. The name Greenpeace was coined in 1970 when a group of 12 Canadians sailed in a 30 year old halibut trawler, the *Phyllis Cormack,* to Alaska's Amchitka Island. The voyage was to protest against United States underground nuclear testing there. The boat didn't reach Amchitka and the bomb did go off, but the voyage aroused such massive public indignation and opposition to the tests that within a year the site was closed and the remaining six tests were cancelled. This was the first of Greenpeace's non-violent direct actions. It was soon followed in 1972 and 1973 by voyages directed against French atmospheric testing in the South Pacific.

From these beginnings Greenpeace has grown to become an international ecological organisation active in the three main areas—conservation, nuclear issues and toxic chemicals—with offices in 15 countries. Once a year representatives from all these offices meet to plan Greenpeace's campaigns for the coming year.

Greenpeace is a totally independent organisation. It is not affiliated to, or supported by, any political organisation, for it believes environmental concerns go beyond party politics. Its international campaigns are funded through contributions from all the Greenpeace offices. This money comes entirely from fund-raising, public donations, memberships and the sale of merchandise.

Non-violent direct action and the Greenpeace ships have, over the years, become the major trademarks: the 11 metre kauri ketch *Vega (Greenpeace III),* built in New Zealand in 1948 and active in the Pacific; the 46 metre *Sirius,* based in Europe; the *Beluga,* a 26 metre riverboat based in West Germany; and the 58 metre *Greenpeace.*

The flagship of the Greenpeace fleet, however, was the *Rainbow Warrior,* until it was sabotaged in Auckland harbour in 1985. The *Rainbow Warrior* was a 44 metre ex-research trawler built in 1955 and bought by Greenpeace in 1977 with a grant from the World Wildlife Fund. In those eight years it carried out campaigns against nuclear reactors; Icelandic, Spanish, Peruvian and Soviet whaling; British, Canadian and Norwegian sealing; nuclear waste dumping and re-processing; nuclear weapons testing; chemical waste dumping; and drift nets. Greenpeace Australia, PO Box 51, East Balmain, NSW 2041; phone: (02) 555 7044.

**green turtle.** Marine reptile belonging to the order Chelonia, family Cheloniidae. The green turtle *Chelonia mydas* may grow to 126 cm in carapace (shell) length and feeds mainly on algae and seagrasses. It takes 50 years before the green turtle reaches a reproductive age. Nesting takes place on certain Great Barrier Reef islands from late October until late March; the eggs hatch about eight weeks later. As the hatchlings take to the sea, predators such as

**A juvenile green turtle *Chelonia mydas* stranded in a coral pool by a fast receding tide.**

seabirds and crabs eat a small percentage though the main predators are fish and sharks, which eat the small turtles as they swim across the reef flat to deeper water. See **turtle** and **carapace**.

**grey nurse shark.** Member of the order Lamniformes, family Odontaspididae. The grey nurse shark *Carcharhinus taurus* grows to about 3.6 metres in length and is probably the most maligned of all sharks. The numbers of grey nurse sharks living in schools on shallow rocky reefs was drastically reduced in the 1970s as these supposed man-eaters were killed by spearfishermen using explosive powerheads on their spearguns. Later research showed them to be a harmless fish-eating species that pose little or no danger to humans unless provoked. They are now a protected species. See **powerhead**, **shark** and **Herbsts nurse shark**.

**The grey nurse shark *Carcharhinus taurus* is a protected species.**

**ground swell.** The background pattern of waves moving across the surface of the ocean. These waves may have been generated by winds thousands of kilometres away.

**gummy shark.** Member of the order Carcharhiniformes, family Triakidae. The gummy shark *Mustelus antarcticus* is recognised by its characteristic teeth which are flat and smooth and arranged in a pavement-like formation in the upper and lower jaws. The gummy shark is bottom dwelling and feeds mainly on crustaceans. It grows to 157 cm in length and is commercially exploited using set and long lines. The flesh is good for eating and it is sold as flake or boneless fish fillets. The levels of heavy metals such as mercury contained in the flesh of these sharks has caused concern in the past. See **shark**.

**The gummy shark *Mustelus antarticus* has plate like jaws containing small teeth which are used for crushing its prey.**

**gunwale.** The top edges of the sides of a ship.

***HABITAT* magazine.** A magazine produced by the Australian Conservation Foundation (ACF) which occasionally has articles on the Great Barrier Reef and environmental issues that might concern scuba divers. Subscription information can be obtained by writing to: ACF, 340 Gore Street, Fitzroy, Vic. 3065. See **Australian Conservation Foundation (ACF)**.

**haemorrhage.** The escape of blood from vessels or arteries caused by trauma or diffusion.

***Halimeda*.** A genus of green algae (division Chlorophyta), common in tropical and subtropical oceans of the world. *Halimeda* species usually have green disc-shaped segments connected by uncalcified flexible nodes. The discs contain large proportions of calcium carbonate in their tissues which on decomposition contributes to the build-up of sediments on coral reefs.

**hammerhead shark.** Member of the order Carcharhiniformes, family Sphyrnidae. There are nine species of hammerhead sharks worldwide and four species are found in Australian waters. All have the characteristically flattened, hammer-shaped head. They are found in all temperate and tropical seas. The hammerhead sharks' diet, ranges from bony fishes to shrimps and crustaceans. Stingrays are also eaten—one hammerhead when caught was found to have 50 stingray barbs embedded in its jaws. All hammerheads bear live young in litters of up to 37 individuals. The scalloped hammerhead, *Sphyrna lewini,* is the most common species worldwide. Large schools of these sharks migrate along the east coast of the United States in summer. Similar migrations do occur in Australian waters, as recently reported by Alby Ziebell, skipper of the dive charter vessel *Coralita,* when diving in the Coral Sea off the Queensland coast. See **shark**.

**handspear.** A hand-held rubber powered spear usually made of stainless steel, aluminium or fibreglass, about two metres long, which is used for killing fish. Handspears are designed so they can be unscrewed into two or three parts for easy transport and storage. The tip has a single point or a cluster of sharp barbs, an explosive powerhead can also be attached via a trumpet shaped adapter which slips over the cluster head. See **gidgee**, **Hawaiian sling** and **powerhead**.

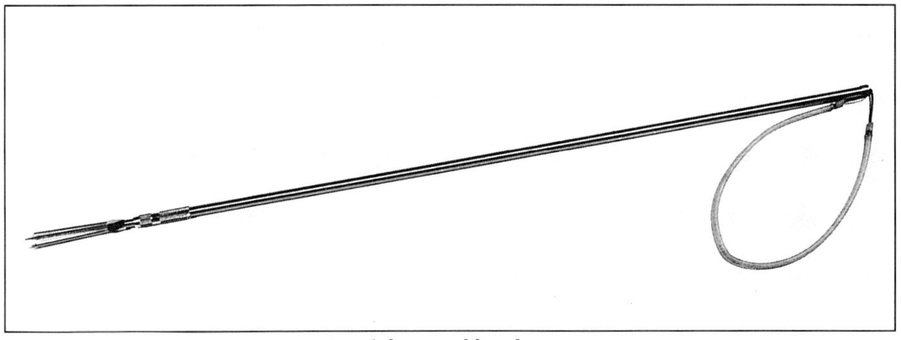

**A stainless steel handspear.**

**hard coral.** Member of the order Scleractinia (the stony corals). There are about 330 species of hard coral on the Great Barrier Reef. All secrete a hard limestone skeleton and each individual polyp has six tentacles or multiples of six surrounding its mouth. Scleractinia can be divided into two main groups; the *hermatypic* or reef-building corals are the most numerous, containing zooxanthellae in their tissues and occurring in warm shallow seas; and the *ahermatypic* or non reef-building corals which lack zooxanthellae and occur in both deep and shallow water environments of many seas. Stony corals secrete a limestone skeleton which provides support and protection for the colony. When coral polyps die their white skeletons remain to form coral rock or be eroded to coral sand.

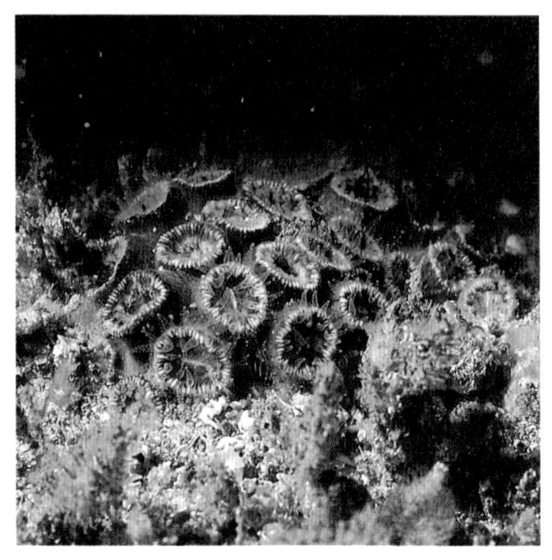

**Hard corals are not restricted to tropical reefs, they also grow in waters off the southern half of Australia, this species *Astrangia woodsi* was photographed in Sydney Harbour.**

**Species of the hard coral *Acropora* vary in shape from erect branching types to flat plates and are often the most conspicuous types of hard corals on some tropical reefs.**

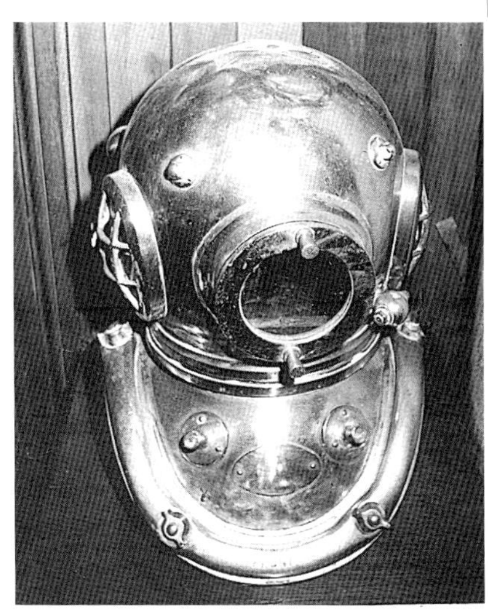

Hard-hats were commonly used for most forms of professional diving in the 1800s and into the mid 1900s.

**hard-hat diving.** Diving with a suit of waterproof material and a helmet with glass viewing windows. Air is pumped continuously from the surface via a flexible hose to the diver's helmet where excess air is bled off via a relief valve. The hard-hat diving suit was manufactured by Augustus Siebe in England in the early 1800s and was used by professional divers worldwide for many tasks including; salvage, maintenance, engineering and pearling operations. More modern forms of diving equipment have rendered the traditional hard-hat equipment obsolete.

**Hass, Dr Hans.** Famous Austrian author, pioneer underwater diver, photographer, film-maker and scientist. Hass started his first exploration dives in the Adriatic in 1938.

**Hawaiian sling.** A hand-held rubber powered spear consisting of an aluminium, plastic or wooden tube with a strong rubber band attached to one end; the device is used for killing fish. The tube is held by the left hand and a thin steel spear is inserted blunt end first and gripped by the thumb and the first two fingers of the right hand, together with the rubber band. The sling is used by stretching the rubber as much as possible and then releasing the grip suddenly, causing the steel spear to shoot forward with great velocity. See **gidgee** and **handspear**.

This carapace of a freshly killed turtle is being prepared for sale and is proudly displayed by a Fijian youth ignorant to the damage caused to the declining turtle populations by continuing this practice.

**hawksbill turtle.** Marine reptile and member of the order Chelonia, family Cheloniidae. The hawksbill turtle, *Eretmochelys imbricata,* grows to about 88 cm in carapace (shell) length and is omnivorous, feeding on algae, ascidians, bryozoans, molluscs, sea-grasses and sponges. In Australian waters it is most abundant in the Torres Strait. It is exploited commercially in Asia where it is much prized for its rich mottled carapace (tortoiseshell) which is used to make carvings, combs, hair ornaments and souvenirs. The smaller turtles are stuffed and sold to tourists while the skins from larger specimens are tanned and used in the manufacture of shoes and handbags; the calipee and meat are used in soups for human consumption. In the past tortoiseshell was widely used for spectacle frames, though today molded plastics have largely replaced it. See **calipee**, **carapace**, **omnivorous** and **turtle**.

**The hawksbill turtle *Eretmochelys imbricata*.**

**heliox.** A mixture of helium (an inert gas) and oxygen, which is used in place of compressed air for deep diving to overcome the problem of nitrogen narcosis. One disadvantage of using helium is that body heat is lost more rapidly, leading to hypothermia. See **hypothermia**.

**helm.** The steering wheel or tiller of a boat.

**Henry's law.** A gas law which states that: *the amount of gas that will dissolve in a given volume of liquid at a constant temperature is directly proportional to the partial pressure of the gas.* In practical terms this means that scuba divers have to use their air decompression tables correctly, if not the dissolved nitrogen in their blood may come out of solution to form bubbles which causes decompression sickness (the bends). See **decompression sickness**.

**Herbsts nurse shark.** Member of the order Lamniformes, family Odontaspididae. Herbsts nurse shark *Odontaspis herbsti* is a bottom dwelling species similar in appearance to the grey nurse shark but inhabiting deeper offshore waters along the edge of the continental shelf. The Herbsts nurse shark is very similar in appearance to the smalltooth sand tiger *Odontaspis ferox* and some scientists argue it may be the same species. See **grey nurse shark** and **shark**.

**hermaphrodite.** An individual that possesses both female (ovarian) and male (testicular) reproductive tissue at the same time. Nudibranchs are true hermaphrodites. See **nudibranch**.

**hermit crab.** Decapod crustacean and member of the infraorder Anomura, superfamily Paguroidea. Hermit crabs are different from other crabs in having a soft and vulnerable abdomen to protect so they live in discarded sea-shells. The abdomen is inserted first and if danger threatens the crabs withdraws into the shell blocking the entrance with its large claws. When a crab outgrows its home it simply discards the shell and moves to a larger one.

**The hermit crab uses a discarded sea-shell to protect its soft abdomen.**

**Heron Island.** A 17 hectare coral cay situated within the Great Barrier Reef Marine Park, 72 km east from Gladstone, Queensland. A world-class dive resort famous for its 'bommie' where scores of fishes can be hand-fed and photographed at close quarters. Non-divers can enjoy a trip in a semi-submersible to get a diver's eye view of the local reefs, or a guided reef walk where they can touch, as well as observe, the reef creatures. In the warmer months from October to March numbers of turtles can be seen nesting and laying their eggs on the beach. See **bommie** and **turtle**.

**Heron Island is a coral cay situated off Gladstone, Queensland on the southern section of the Great Barrier Reef complex and is famous worldwide as a scuba diving resort.**

**high-altitude dive.** Any dive conducted at an elevation greater than 304 metres or 999 feet above sea-level, requiring corrections in the dive tables for depth.

**high tide.** The maximum height reached by a rising tide. This level varies throughout the year in accordance with the gravitational pull of the Sun and the Moon. Also called high water.

**high water.** See **high tide**.

**holdfast.** Finger-like projections at the base of seaweeds used for attachment to rocks or other substrates on the ocean floor. This anchors the seaweed and prevents dislodgement in rough seas.

**Holdfasts of the brown alga *Ecklonia radiata* are home to a myriad of plants and animals including: algae, bryozoans, tubeworms, molluscs, and small sponges.**

**holothurian.** See **sea-cucumber**.

**holotype.** The specimen from which the description of a new species is made, usually held in a museum or university.

**hood.** The neoprene headpiece of a wetsuit which fits over a diver's head and prevents heat loss to the surrounding water. The hood can be worn separately or can be permanently sewn and glued to the main body of the wetsuit.

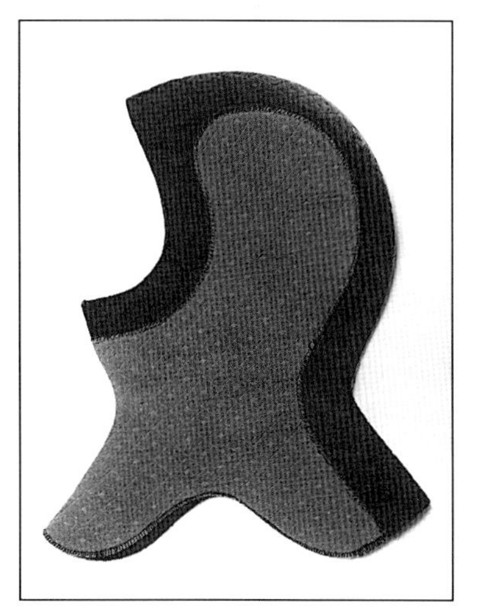

**A large proportion of a diver's body heat is lost through the head. Thus it is essential for safe diving to wear a properly fitting neoprene hood when diving in cold water to avoid hypothermia.**

**hookah.** A portable surface-supplied breathing apparatus which consists of a fuel driven (or electric) low pressure compressor and reserve air chamber with about 100 metres of hose attached to a demand valve (regulator). Depending upon the power of the motor driving the compressor a hookah may support from one to four divers at depths of up to 35 metres. All physiological problems associated with scuba diving are relevant to hookah divers. It is essential that proper maintenance be carried out on hookah equipment to prevent carbon monoxide poisoning. These units are commonly used by professional abalone divers and other commercial divers. See **compressor**.

**horseshoe crab.** See **Xiphosura**.

**humpback whale.** Baleen whale and member of the order Cetacea, sub-order Mysticeti (baleen whales), family Stenidae. Baleen whales are so called because their mouths contain rows of fringed plates called 'whalebone' or 'baleen'. The humpback whale *Megaptera novaeangliae* has a distinctive curved back, very long white coloured flippers (up to four metres), and many white bumps consisting mainly of barnacles, on the head and body. Humpbacks grow to about 15 metres in length and 50 tonnes in weight. All whales make underwater sounds but the greatest of all, the real opera star of the oceans, is the humpback. Its song is more complex, longer and more varied than that of any other whale. Humpback whales travel in small family groups at speeds of up to four knots and migrate along the Australian coast each year. They are a common site in the Ningaloo Marine Park off the North West Cape of Western Australia. In June and August they are common off the New South Wales and Queensland coasts. Good observation points are Cape Byron in New South Wales and Moreton Island in Queensland. The International Whaling

Commission banned the taking of humpback whales in waters south of the equator in 1963. See **cetacean** and **whale**.

**Humpback whale sightings are becoming more common all around the Australian coastline. This mother and calf was sighted off Broome, Western Australia.**

**hurricane.** Winds in excess of 64 knots or 118 km/hr.

**hydrocoral.** Stinging hydrozoan coral and member of the class Hydrozoa, families Milleporidae and Stylasteridae. Hydrozoan corals in the family Milleporidae are usually green, cream or yellow in colour and grow as branching, encrusting and vertical plate forms in tropical waters. They are commonly called fire corals or stinging corals because they often cause a burning sensation when they come in contact with softer types of human skin and can cause a nettle-like rash which may last up to 10 days. Hydrozoan corals in the family Stylasteridae (*Stylaster* spp. and *Distichopora* spp.) can be found under ledges and on the roofs of caves worldwide, extending to the Arctic and Antarctic regions. They are conspicuous for their delicate fan-shaped branching and their bright pink, yellow and purple colour variations. These colours remain after the polyps have died.

**The fire-coral *Millipora platyphella* can cause a burning sensation when touched as can all species in the genus *Millipora.***

**The mauve hydrozoan coral *Distichopora* sp. is common under ledges and on the roofs of caves in tropical waters.**

**hydroid.** Stinging invertebrate and member of the class Hydrozoa, order Hydroida. Hydroids have delicate, feather, fern or flower-shaped colonies which contain nematocysts capable of causing severe irritation to the unprotected skin of scuba divers. When divng you should wear a lycra suit or wetsuit (depending on the water temperature) for protection from these very common stinging organisms. Hydroids are found at all depths from the intertidal zone to the deepest abyss and are particularly numerous in the waters of southern Australia, especially in areas with fast flowing currents such as tidal channels.

**Feathery hydroids can cause a nettle-like sting when touched by soft or unprotected human skin.**

**Hydroids come in a variety of shapes. This species *Ralpharia magnifica* is stalked and has a flower-shaped body.**

**Hydro-Jet.** A revolutionary new scuba propulsion unit which operates on air. The Hydro-Jet has a small five-cylinder engine that runs off a 80 cubic foot, 3000 psi scuba cylinder. The engine is attached to the back of the scuba cylinder and the cylinder is worn in the normal position. This unit provides a hands-free means of forward propulsion. This system is unique for a number of reasons, the most innovative feature probably being the recycling of air from the engine into a modified regulator mouthpiece from which the diver breathes. This information was supplied by the Beyond International Group, Sydney. The inventor, Robert Hyde, can be contacted at: Hyde Power Systems, 9340 West Putter Court, Crystal River, Florida 32629, USA; phone: (904) 795 3340.

**hydrozoan.** Member of the class Hydrozoa. See **hydrocoral** and **hydroid**.

**hyper-.** A prefix meaning greater or larger than.

**hyperbaric.** Increasing ambient pressure.

**hyperventilation.** Increased or forced respiration causing a purging of carbon dioxide from the lungs and suppressing the desire to breathe. This technique can extend the bottom time of breathhold divers but unconsciousness can occur without warning. The brain regulates breathing based on carbon dioxide levels in the blood. Breathing heavily to rid the lungs of carbon dioxide before diving can delay the natural instinct to breathe and can have tragic consequences. As the breathhold diver descends, the ambient pressure causes increases in the partial pressure of the oxygen and carbon dioxide in the lungs and a decrease in the volume of these gases, causing the oxygen to dissolve more readily into the bloodstream. On ascent the diver's oxygen partial pressure falls and a sudden lack of oxygen (anoxia) could cause him/her to lose consciousness without warning. Anoxia has been responsible for the deaths of many spearfishermen over the years, especially when hyperventilation was used to extend the underwater hunting time.

**hypo-.** A prefix meaning smaller or less than.

**hypobaric.** A decrease in pressure from atmospheric to subatmospheric levels.

**hypothermia.** The loss of body heat to the environment causing subnormal body temperatures. When exposed to the ocean environment scuba divers loose heat to the water continuously and hypothermia can result. Hypothermia occurs in water because body heat is lost some 25 times faster in water than in air. The lethal lower core temperature limit for humans is about 23°C–25°C (rectal temperature). Never give alcohol to a diver with hypothermia as the surface blood vessels dilate and fill with blood from the deeper warmer tissues; this causes an overall decrease in temperature which could be fatal to the patient. When diving, match the thickness of your wetsuit with the temperature of the water and when you start to show the first signs of hypothermia, such as shivering, be aware you might have to finish your dive earlier. First aid measures include:

• Keeping the patient out of the wind and as warm and dry as possible using any material at hand such as towels, blankets or tarpaulins. If possible have someone snuggle up to the patient to help restore lost body heat.
• If conscious give warm drinks with plenty of sugar.
• Give mouth to mouth resuscitation and external cardiac massage if the patient's heart and breathing have ceased.
• As soon as possible seek medical help; if no help is available place the patient in a warm bath or shower at 36°C and raise the temperature over a ten to fifteen minute period to about 42°C. The limbs should be elevated and held clear of the water to prevent hypotension. See **space blanket**.

**hypoxia.** See **anoxia**.

**ichthyology.** The study of fishes, their anatomy, life history, and classification. A Swede, Peter Artedi, is generally considered to be the father of ichthyology and his main work was published in 1738; he proposed a systematic division of fishes into orders, genera, species and varieties.

**Indo-Pacific Marine Aquarium.** An aquarium complex built in Darwin, featuring fishes, corals and anemones. Each aquarium is a self-contained eco-system, the only introduced food being an occasional fish for the carnivorous species. Located at: Lambell House, Smith Street, West Darwin, NT 5794; phone: (089) 811 294.

**inflatable boat.** A type of boat used by divers, also known as a 'rubber duck'. Some of the popular brands include; Achilles, Avon, Lancer, Metzler and Zodiac. These boats are lightweight, buoyant, collapsible and take approximately 20–30 minutes to assemble. They are quite expensive but are very seaworthy and can carry greater loads than conventional boats of the same size. Small inflatables are often used as tenders to larger boats; being made of a rubberised nylon or a PVC type plastic polymer, they do not damage the mother boat when docking.

**Inflatables make very good dive boats. They are easy to transport and can carry heavy loads which conventional boats of the same size could not handle.**

**inflatable marine rescue tube.** See **Safety Sausage**.

**inhalation.** The breathing in of air or other gases.

**inlet.** A small opening on a shoreline.

**inspiration.** The act of breathing in.

**insurance for divers.** A contract to guarantee against risk or harm especially designed for divers. Professional Indemnity Insurance, Personal Accident Insurance and Dive Club Public Liability Insurance is available through the Australian Underwater Federation, Box 1006, Civic Square Canberra ACT 2608; phone: (06) 247 5554.

**International Shark Attack File.** See **Australian Shark Attack File**.

**intertidal zone.** A zone of rock, sand or mud between low and high tide marks. This area is the home of many species of marine plants and animals.

**invertebrate.** An animal without a backbone. Over 90% of the animal species living in the oceans are invertebrates and they occupy every habitat from above the high tide level to the abyssal depths of the oceans.

**Irish moss.** Red alga and member of the division Rhodophyta. Irish moss *Chondrus crispus* is a purple-brown coloured seaweed sometimes called carragheen, which grows mainly in the waters off the Atlantic coasts of America and Ireland. After harvesting and processing it is often used as a substitute for gelatine, and as a thickening agent in food and pharmaceutical products.

**isinglass.** A gelatine-like substance found in the bladders of fishes, such as the European sturgeon, used in brewing beer and as an adhesive and preservative.

**isopod.** Crustacean and member of the class Malacostraca, order Isopoda. Most marine isopods have a flattened body, are well camouflaged and have a leathery external skeleton made up of a series of plates, much like an armadillo in appearance. They vary in size between one and eight millimetres in length and live in soft sediments from the shallows down to 10000 metres. Their feeding habits vary, from eating organic debris and algae to boring into marine timbers and cables and parasitising marine animals such as fish and decapod crustaceans. See **ectoparasite** and **fish-lice**.

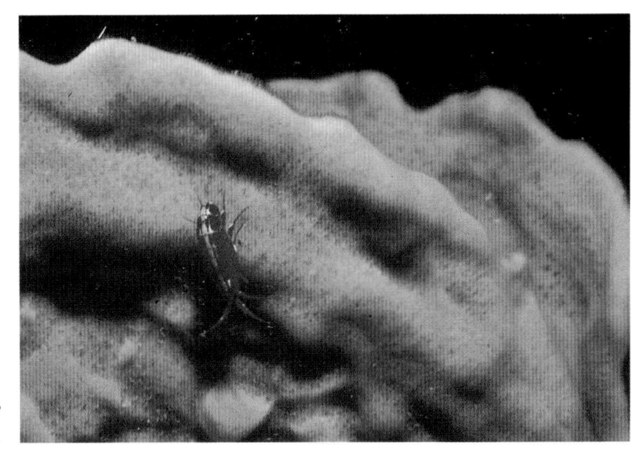

**An isopod scavenging for organic debris on a sponge.**

***Janthina* spp.** Floating predatory violet snails and members of the family Janthinidae. This mollusc produces a float by secreting a foam-like mass of mucus which entangles bubbles of air and then hardens. The mollusc drifts with the wind and the tide and feeds upon other planktonic animals. There are a number of species and they are found in the Pacific, Atlantic and Indian oceans.

**Japanese oyster.** See **Pacific oyster**.

**jellyfish.** Coelenterate and member of the phylum Cnidaria, class Scyphozoa. Jellyfish are often seen swimming just below the surface of the sea and are generally bell-shaped with long trailing tentacles. They move by contractions of muscles on the outer rim. The tentacles are covered with stinging cells (nematocysts) that are used to capture food which is returned to the mouth parts under the bell for digestion. See **Catostylus mosaicus**.

**jetsam.** Material thrown overboard to lighten a vessel in distress, sometimes finding its way ashore, but usually sinking. See **flotsam** and **ligan**.

**jetty.** A structure such as a pier or wharf which extends from the shoreline into a lake, river or ocean and is used for the loading and unloading of passengers or goods from boats.

**jewel anemone.** A colonial anemone and member of the subclass Zoantharia, order Corallimorpharia. The jewel anemone varies in colour from red and orange to pink with non-retractile white-knobbed tentacles and occurs in caves and under overhangs in shallow water off southern Australia. See **Zoantharia**.

**The jewel anemone *Cornyactis australis* is easily identified by its tentacles, which are short and end in a swollen white knob.**

**J-valve.** A valve which attaches to a scuba cylinder and allows a measure of air to be held in reserve until required. Also known as a J-reserve or a constant reserve valve. It has a J-shaped rod attached to a lever on the valve stem which allows the reserve mechanism to be pulled on or off. The J-valve is similar in construction to the K-valve except it has a spring which holds a teflon block in place until the pressure in the cylinder drops below about 20 atmospheres. At this point the spring isolates the remainder of the cylinder contents until the J-valve is mechanically held open by pulling on the attached rod. Only then is the remaining 20 atmospheres released to the diver via a regulator. The modern contents gauge has replaced the J-valve because it enables the scuba diver to accurately monitor air supply throughout the dive. See **K-valve**.

**The J-valve is also known as a constant reserve valve.**

**kakka.** See **cacker**.

**kelp.** Large brown algae and members of the division Phaeophyta. The most common species in Australia are: *Ecklonia radiata* (east coast), *Macrocystis angustiflora* (south-eastern Australia) and *Durvillaea potatorum* or bull-kelp (southern Australia). Kelp forests are home to a miriad of plants and animals from tiny bacteria and fungi which live in the surface slime of the kelp fronds to the many encrusting organisms such as bryozoans, hydroids, tunicates and seaweeds which provide food for many species of fish, crabs, shrimps and molluscs. See **brown alga**.

**The brown kelp *Ecklonia radiata*.**

**The brown kelp *Macrocystis angustiflora*, has gas-filled bladders which buoy the large fronds at the surface.**

**The bull kelp *Durvillaea potatorum* grows to over seven metres in length and also occurs in waters off South America.**

**key.** See **coral cay**.

**killer whale.** See **orca**.

**king crab.** See **Xiphosura**.

**KMB.** A designation of the Kirby-Morgan Band mask used by commercial divers, for mixed gas, surface supplied air and in some instances scuba diving.

**knee guards.** Neoprene knee guards or special knee pads sewn and glued onto a wetsuit to protect it. Probably the most vulnerable part of any wetsuit is the knee area and this is the first area to show signs of wear from contact with the bottom or other objects, especially hard coral.

**knot.** A speed of one nautical mile/hour. In other units it is expressed as approximately: 1.85 km/h, 1.7 feet/second or 0.5 metres/second.

**krill.** Shrimp-like crustaceans of the order Euphausiacea that occur in giant schools in the Arctic and Antarctic oceans. Whales are one of the main predators of krill and as whales have been killed in large numbers over the years this has caused an increase in krill mass. However even the krill may be at risk in the years to come: Soviet and Japanese trawlers have been harvesting krill for use as stock feed. If profitable, krill could be harvested by many nations in the future; what result this may have on the ecology of the region is unknown.

**K-valve.** A simple on-off tap which is screwed onto a scuba cylinder to allow a regulator (which regulates the flow of air) to be mechanically attached to the cylinder. The K-valve is also called a pillar valve and is now the most common valve attached to scuba cylinders. See **J-valve**.

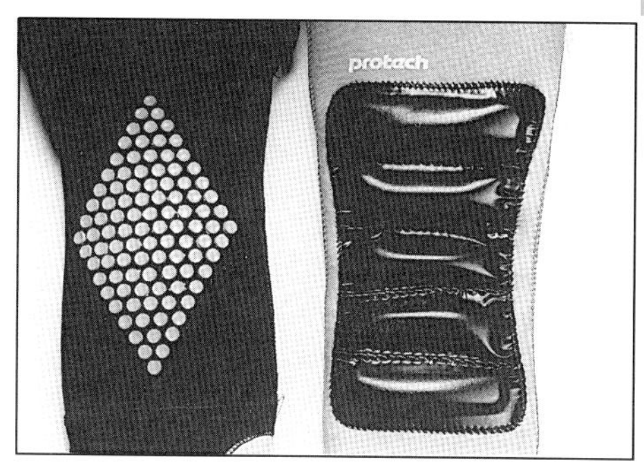


**Wetsuit knee guard design varies with different manufacturers.**

**Krill, a collective term for a number of species of shrimps belonging to the family Euphausiidae.**

**A K-valve is also called a pillar valve and is the type of valve fitted to most scuba cylinders.**

**lanyard.** A line, rope or cord used for securing something, especially on a boat.

**lateral line.** A mark or line running the length of both sides of a fish's body. It carries sensitive receptor cells that can pick up vibrations caused by sound and water pressure variations. The lateral line enables a fish to be aware of its environment, including the presence of prey and predators.

**The lateral line runs along both sides of a fish's body and keeps it informed via sensory receptors of changes in its environment.**

**learning to dive course.** A method by which a person learns to scuba dive—gaining a certificate of competency called a 'C' card. In order to learn how to scuba dive the prospective diver should be able to swim and float, be free of diabetes, epilepsy, asthma or any serious heart or lung disease. The candidate should also be reasonably fit, not grossly overweight and be capable of equalising pressure in the ears and sinus passages on descent. Dive store managers can provide the name of a diving doctor and a diving medical examination is required prior to commencement of a dive course. If undertaking a scuba course while on holiday, a recognised NASDS, NAUI, PADI or SSI course is the only safe way to learn. Some holiday dive resorts issue their own certification which is not recognised elsewhere. See **'C' card**, **BSAC**, **NASDS**, **NAUI**, **PADI** and **SSI**.

**leatherback turtle.** Marine reptile and member of the order Chelonia, family Dermochelyidae. The leatherback turtle *Demochelys coriacea,* also called the leathery turtle, is the world's largest species of turtle and is endangered due mainly to over exploitation of its eggs which are collected from beach nesting sites in certain Asian countries and sold as food. The leatherback turtle has five distinct ridges and no scales on the carapace (shell) and may measure up to two metres in carapace length. It is widespread throughout tropical and temperate seas. It is reported to feed on jellyfish and salps and due to its selective diet this turtle is particularly susceptible to premature death due to ingestion of plastic debris in the water. See **turtle** and **carapace**.

**lee.** A sheltered place away from the wind, or on the protected side of a ship.

**leech.** See **marine leech**.

**leeward.** Away from the wind; the sheltered side; or the point or quarter towards which the wind blows.

**leeway.** The distance a ship is forced sideways from her course by a wind.

**leopard shark.** See **zebra shark**.

**lift bag.** A bag constructed from heavy duty material, which is used underwater to salvage heavy objects. The lift bag has an open end into which air is pumped either from a regulator mouthpiece or from a surface supply, the amount of air varying with the degree of lift required. The bag may have an air venting device to help control ascent rate of the object being lifted. Lift bags come in various sizes suitable for lifting small anchors to large boats.

**Lift bags are used to help recover heavy objects from the sea-floor and are inflated via a regulator mouthpiece or auxiliary air supply.**

**ligan.** Any object that has sunk but has been marked with a buoy so it can be relocated. See **flotsam** and **jetsam**.

**light breeze.** A wind of 4-6 knots or 7-11 km/hr.

**light penetration with increasing depth.** The selective filtration of sunlight in water. When sunlight travels through water selective filtration of the various wavelengths takes place; 0-5 metres red and orange light is absorbed; 5-10 metres yellow light is absorbed; 10-20 metres the remaining colours are blue and indigo; at depths greater than 33 metres the human eye is unable to distinguish individual colours other than blue.

**lightstick.** See **Cyalume**.

**limnology.** The study of lakes and ponds.

**littoral zone.** The seashore can be divided into zones characterised by the various plant and animal associations. The highest zone is the supralittoral zone (spray zone) which is above high tide level, the eulittoral zone (main part of the seashore) is the zone between the high and the low tide levels, the lowest zone is called the sublittoral zone (subtidal zone). The animals and plants found in these zones vary greatly with the type of coastline and the latitude.

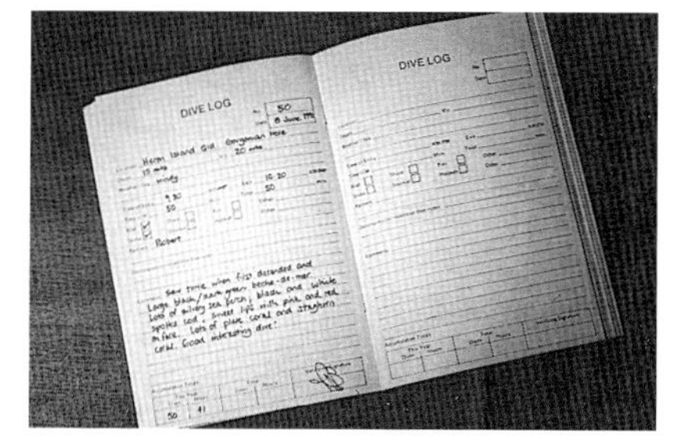

**Dive logbooks should be filled out after every dive.**

**logbook.** A book in which scuba divers record details of their dives. Scuba divers should record basic information about every dive in their logbook: name of buddy, depth, time, duration and any interesting observations. This book provides a permanent record of a diver's history and can be produced to prove his/her experience level to a dive organiser.

**loggerhead turtle.** Marine reptile and member of the order Chelonia, family Cheloniidae. The loggerhead turtle, *Caretta caretta,* may reach 110 cm in carapace (shell) length and occurs in all tropical and warm temperate oceans where it feeds on a diet of crustaceans, fish, jellyfish and molluscs. The main breeding site on the Great Barrier Reef is Wreck Island (Capricorn/Bunker Group) where up to 1000 females nest each season. See **turtle** and **carapace**.

**The loggerhead turtle *Caretta caretta* includes jellyfish in its diet.**

**Lord Howe Island.** A sub-tropical island situated due east of Port Macquarie on the New South Wales coast, 700 km northeast of Sydney. Lord Howe Island has been included on the World Heritage List, making it one of only three islands in the world to be so honoured. The island is surrounded by the southern-most coral reefs in the world. Its location, in an area where temperate and tropical ocean currents meet, contributes to an unusually diverse and sometimes unique population of marine plants and animals—a popular spot for scuba diving. Elizabeth and Middleton Reefs lie 95 km to the north and together with the reefs off Lord Howe Island are home to about 111 species of hard coral, 60 species of crustaceans and 200 species of fishes, some of which are endemic to the area. In recognition of their importance Elizabeth and Middleton Reefs have been proclaimed Marine Nature Reserves (1987).

**low tide.** The lowest level reached by a falling tide.

**low water.** See **low tide**.

**luminescence.** See **bioluminescence**.

**lycra suit.** A thin stretchy suit made from lycra, used mainly for tropical diving to protect against hydroids, corals, jellyfishes and other marine hazards. In southern waters lycra suits are worn as an undersuit beneath a conventional wetsuit. They provide extra warmth and make getting in and out of wetsuits easier. Underwater photographers often dress their models in brightly coloured lycra suits to give extra colour to photographs. New suits containing a mixture of lycra and nylon provide more insulation and protection than lycra alone.

**This trio of divers are each wearing a lycra suit, ideal swimwear for warm tropical waters.**

**magnetometer.** An instrument for measuring the direction and intensity of magnetic forces. Scuba divers may use hand-held models to detect buried or encrusted metal objects and larger versions may be towed by boats to detect shipwrecks and lost aircraft.

**mako shark.** Member of the order Lamniformes, family Lamnidae. There are two types of mako sharks, *Isurus oxyrinchus* the shortfin mako, also known as the blue pointer and *Isurus paucus* the longfin mako. Mako sharks can grow to four metres in length and to over 500 kg in weight, feeding mainly on fishes and squid. They are prized sporting fish renowned for jumping up to six metres out of the water when hooked. See **shark**.

**Malacological Society of Australia.** The following information has been supplied by the Malacological Society of Australia:

A Society that was established in 1955 to foster the study of molluscs and to encourage environmentally responsible shell-collecting in Australia. The Society has several branches throughout Australia which hold regular meetings for talks and discussions on shells and which arrange local excursions.

Members of the Society receive a quarterly newsletter *Australian Shell News* and the annual Journal. The newsletter includes articles on shells, reviews of the literature, exchange requests and advertisements by shell dealers.

The Journal presents scientific research, the results of local and overseas molluscan workers. Enquiries should be directed to: Hon. Secretary Malacological Society of Australia c/- Invertebrate Zoology Dept. Museum of Victoria, 285 Russell Street, Melbourne Vic. 3000.

**Manly Oceanarium.** A circular three level oceanarium complex in Sydney, built on the site of the old Marineland at a cost of $13 million. The main aquarium tank holds four million litres of sea water and features a variety of sharks and a large 300 kg groper. The fishlife can be observed from a moving walkway enclosed in an acrylic dome which passes through the main tank. Other smaller displays include an eel tank, living coral tank, a crustacean tank and a Sydney Harbour tank. There is also an audio-visual display and the Discovery Room where marine life educators are available to explain the underwater world and answer your questions. On the top level of the complex there are Australian and New Zealand fur seals to entertain the children. Manly Oceanarium, West Esplanade, Manly, NSW 2095; phone: (02) 949 2644.

**mantle.** In molluscs the soft tissue which secretes the shell.

**mariculture.** The technique of artificially cultivating marine plants and animals for human consumption and stock feed. The milkfish *Chanos chanos*, have been farmed in Indonesia since the 15th century and more recently in the Philippines and Taiwan. In some experiments using human waste as fertiliser, yields of fish up to 500 tonnes per square kilometre have been attained. Fish-farming may someday replace traditional fishing and allow natural

fish stocks to regenerate, saving certain species from extinction by overfishing. See **aquaculture**.

**marine algae.** The marine algae fall into three main groups or divisions. The colour of the alga in most cases determines the division into which it is placed: red algae (Rhodophyta), green algae (Chlorophyta), and brown algae (Phaeophyta).

**marine angiosperm.** See **sea-grass**.

**marine aquaria public.** Locations where marine plants and animals are on display for the public. See individual aquaria:
- Great Barrier Reef Wonderland, Townsville Qld.
- Indo-Pacific Marine Aquarium, Darwin NT.
- Manly Oceanarium, Manly NSW.
- Sydney Aquarium, Darling Harbour, Sydney NSW.
- Underwater World, Perth WA.
- Taronga Zoo Aquarium, Sydney NSW.

**Marine Education Society of Australasia (MESA).** A society which offers conferences, professional training activities, overseas study tours and newsletter to encourage wise use, understanding and enjoyment of our marine environment. Secretary of MESA, PO Box 1379, Townsville, Queensland 4810.

**marine leech.** Annelid worm and member of the class Hirudinea. Marine leeches have a large sucker at the posterior end and a smaller sucker at the anterior end surrounding the mouth. Leeches obtain their food by attaching to a host animal with their mouth parts and sucking their blood and body fluids. Some species are host-specific while others infest many animals, from seabirds to oysters and fishes. See **ectoparasite**, **numbfish** and **spiracle**.

**These three marine leeches are sucking body fluids from a short-tailed electric ray and have attached themselves close to a spiracle just behind one of the eyes.**

**Marine Life Resources Kit.** A kit that contains marine life posters, teacher's guide to marine life, activities guidelines and worksheets. A valuable aid to marine educators and school teachers. Available from: Marine Studies Centre, PO Box 138, Queenscliff, Vic. 3225; phone: (052) 523 344.

**marine parks.** See **Great Barrier Reef**, **Marmion Marine Park**, **Ningaloo Marine Park** and **Lord Howe Island**.

**marine research stations.** Sites where marine research is carried out in Australia.

• The Marine Studies Centre, built over the sheltered waters of the D'Entrecasteaux Channel south of Hobart and operated by the Tasmanian Education Department. This marine centre has for many years offered students of all ages 'hands on' experience of the marine environment. Up to 6000 students a year visit the centre. For further information contact: Marine Studies Centre, Jetty Road, Woodbridge, Tas. 7162; phone: (002) 67 4649.

• The Waterman Marine Research Laboratory, is operated by the Fisheries Department of Western Australia and is adjacent to the Waterman Reef Observation Area. This reef has been kept in a condition as close as possible to its virgin state by prohibiting the taking of marine plants or animals. Waterman Marine Research Laboratory, West Coast Highway, Perth, WA; phone: (09) 447 1366.

• Research stations on the Great Barrier Reef can be found on: Lizard Island (operated by the Australian Museum); One Tree Island, near Heron Island (operated by the University of Sydney); Orpheus Island (operated by James Cook University); and Heron Island (operated by the University of Queensland).

• The Western Australian Museum has a research station on Beacon Island in the Wallaby group of islands, Western Australia.

• Queenscliff Marine Station, situated on the Bellarine Peninsula, near Port Phillip Heads in Victoria is owned and operated by the Victorian Institute of Marine Sciences (VIMS), in collaboration with Monash University, Royal Melbourne Institute of Technology and the University of Melbourne. The Marine Station is home to the Deakin University aquaculture laboratory at Point Henry, the Victorian Government's Marine Science Laboratories, and VIMS Marine Studies Centre, which offers public education programmes about the sea and its resources. For further information contact: Queenscliff Marine Station, Victorian Institute of Marine Sciences, PO Box 138, Queenscliff, Vic. 3225.

**Marmion Marine Park.** A Western Australian marine park declared in 1987, for public recreation and the conservation of marine plants and animals, situated just north of Perth. The park extends along the coastline from Trigg Island to Burns Rock and five kilometres seaward. Scuba diving is excellent on the extensive offshore limestone reefs which have many caves and ledges that provide hiding places and homes for a variety of marine life. An underwater dive trail is being established on Boyinaboat Reef. This reef is only 50 metres from the southwest wall of Hillarys Boat Harbour at Sorrento. In the middle of the Park, at Little Island, divers may be lucky enough to have an underwater encounter with an Australian sea lion. Western Australia has more declared historic shipwrecks than any other State or Territory, and one of these, the *Centaur*, lies on Centaur Reef in the southern section of the park. For more information on this area contact: Marmion Marine Park Manager, Department of Conservation and Land Management, Hillarys Boat Harbour, Hillarys, WA 6025; phone: (09) 367 0333.

**mask.** See **face mask**.

**mask clearing.** Technique for clearing water from a face mask while still submerged. This can be achieved by exhaling through the nostrils while tilting the head back and holding the top edge of the mask onto the face with the first two fingers. A number of variations of this technique are also commonly used. See **mask squeeze**.

**mask squeeze.** Reduced pressure in the air space of a face mask on descent, due to the pressure not being equalised. If the pressure difference is excessive the facial tissues will swell and/or small blood vessels in the eye will rupture. Mask squeeze can be eliminated by simply exhaling through the nose into the mask. See **mask clearing**.

**Mayday.** An international radio distress signal used by ships and aircraft. A **Mayday** situation is one in which you are in **grave** and **imminent danger** and you require immediate assistance; your boat is sinking or is badly holed (e.g. on a reef in bad weather). A **pan** situation on the other hand is one involving a ship or person, where there is no immediate danger; non-urgent medical assistance may be needed, the boat may have run out of fuel and be drifting, but danger is not imminent. Penalties for sending false distress signals are severe.

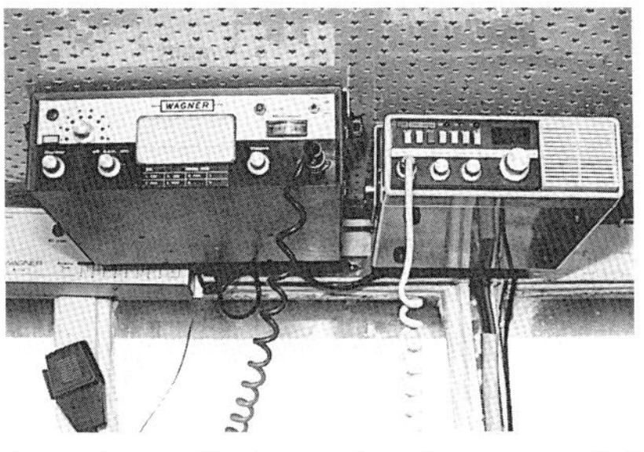


**A marine radio transceiver is an essential item of safety equipment for boat owners.**

How to send a distress call

Using one of the recognised international radio distress frequencies, transmit the following:

• **Mayday Mayday Mayday**.
• This is ................ (name and radio call sign of your vessel three times).
• **Mayday** ............. (name and radio call sign).
• Your position.
• The cause of the distress and the help required.

How to acknowledge a distress call

On the same frequency as the distress call, transmit the following:

• Name and call sign of vessel in distress three times.
• This is .... (your vessel's name and call sign three times).
• Received **Mayday**.

If no other vessel has acknowledged the **Mayday**, or if you are nearer or better suited to assist than another vessel, you should transmit the following information, in the order shown, and on the same distress frequency:

• Name of your vessel.
• Your position.
• Your speed and the approximate time it will take you to reach the vessel in distress. See **radio distress frequencies**.

**Mayoe, Jacques.** French-born breathhold diver who attained the breathhold world record depth of 105 metres (344 feet) off Elba, Italy, in 1983. This record stood until 1989. See **breathhold deep diving record**.

**mermaid.** A mythical marine creature with the head and shoulders of a woman and the tail of a fish. The image of pregnant and suckling female dugongs exposing their enlarged breasts when surfacing to breathe, could in the eyes and mind of the lonely sailor give rise to the mermaid myth. A grandmother's tale of such a creature which was supposedly caught in fishing nets, after a storm on a beach near Bristol, England, sometime prior to the First World War, is intriguing. The remains of the creature was supposedly held by the Bristol Museum.

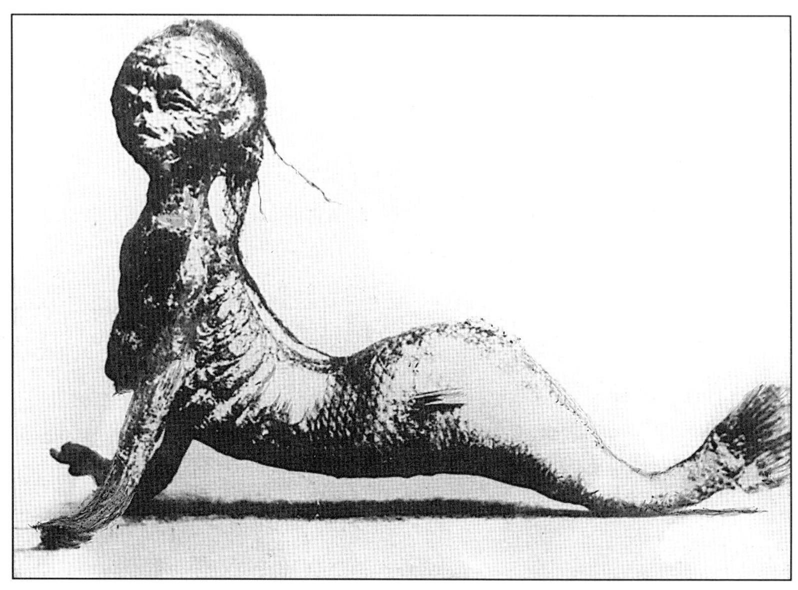

**This mermaid was supposedly washed up on a beach near Bristol, England sometime prior to the First World War. It was common for sailors and showmen of the time to manufacture mermaids from various animal body parts.**

**mermaid line.** See **float line**.

**metal detector.** See **underwater metal detector**.

**midden.** A mound consisting of shells of edible molluscs and other items such as charcoal from cooking fires left by prehistoric man. Many middens are found on the Australian coastline, as Aborigines in the past have often feasted on the plentiful

**This midden in the sand dunes at Port Stephens, NSW consists mainly of the pipi shell *Donax deltoides*.**

supplies of shellfish (such as pipis and many varieties of oysters and whelks) found on our sandy beaches and marine rock platforms. See **pipi**.

**middle ear barotrauma.** The most common diving related medical disorder suffered by scuba divers. The primary cause of middle ear barotrauma is the failure of divers to equalise pressure in the middle ear during descent, resulting in a burst eardrum. On ascent a reverse blockage can occur especially when the effects of a decongestant have worn off during a dive, resulting in tissues swelling around the eustachian tube, ultimately leading to the eardrum bursting outwards. Smoking is another contributing factor as all smokers have 'boggy' mucus membranes around the nose and eustachian tubes, which can lead to blockages on descent or ascent and cause barotrauma. To prevent middle ear barotrauma make sure equalisation occurs before pressure is felt in the ears and do not dive with a cold, allergies or after taking decongestants especially if you are a heavy smoker. Finally if your eardrum is damaged do not try to clear your ears in the usual manner as this will cause further damage. See **barotrauma**.

**mixed gas diving.** Use of a mixture of gases for deep diving. Professional divers are sometimes required to dive to great depths, well outside the depth capabilities of the sport diver. To achieve these depths safely a mixture of gases is used, usually a helium/oxygen combination which eliminates the dangerous build-up of nitrogen in the tissues that would occur if normal compressed air was breathed at depth. This combination also prevents oxygen toxicity where depths beyond 90 metres are involved, as the oxygen content of the mix can be varied to suit the depth. See **oxygen toxicity** and **partial pressure**.

**moderate breeze.** A wind of 11-16 knots or 20-30 km/hr.

**mollusc.** A member of the phylum Mollusca. Molluscs are mainly aquatic animals and include five main classes:
1. Polyplacophora. The chitons or coat-of-mail shells have a shell consisting of eight overlaping plates.
2. Gastropoda. The univalve shells such as cowries, snails, limpets, whelks and shell-less slugs called nudibranchs.
3. Bivalvia. Mostly bottom-living animals such as clams, mussels, oysters and scallops which have two calcareous halves, hinged at their dorsal margin and form the typical bivalved shell.
4. Scaphopoda. The elephant-tusk shells which have a curved tapered tube, open at both ends. They burrow into soft sediments.
5. Cephalopoda. The highly developed molluscs such as the cuttlefish, octopus, squid and nautilus.

Molluscs are abundant in all seas and the gastropod molluscs are very common in the intertidal zone, where they feed on algae, fungi and cyanobacteria. They use their tongue-like radula as a scraping tool to remove the food from the substrate. Some species are carnivorous whereas others lack a radula and obtain their food by filtering the sea water. Shell collecting is a profitable business, as humans have long been fascinated by intricate and beautiful seashells and have found many uses for them, including: buttons, ornaments and jewellery. In the past, shells were commonly used as a form of money by island peoples and the practice is still carried out today on some isolated Pacific islands. The popularity of scuba diving has made possible the collection of shells from areas where dredging is impossible and this has resulted in divers finding shells that were once thought to be very rare but in reality are quite common in localised areas. Molluscs such as squid and octopuses are heavily fished in some areas for human consumption. See **cephalopod**, **clam**, **cone shell**, **cowrie**, **nautilus**, **nudibranch**, **oyster** and **pipi**.

**Monkey Mia.** World famous caravan park in Shark Bay, near the township of Denham, Western Australia, where wild dolphins come in from the Indian Ocean every day to be handfed and petted by humans. Tourists should visit the Dolphin Information Centre to ensure they know how to treat the dolphins before they meet them. Stand knee deep in the water and wait for the dolphins to approach, stroke the sides of the dolphins but don't touch the head (particularly near the blowhole), the dorsal fin, the flippers or the tail. The Dolphin Information Centre can be contacted by phone on: (099) 481 366 and has displays, videos, books and information sheets and sells fish for feeding to the dolphins. See **dolphin**.

**mooring.** A place where a vessel can or may be moored by attaching cables or lines to a solid object.

**mouth brooder.** Fish species which incubate fertilised eggs in the mouth prior to hatching. Common marine mouth brooders include the cardinalfish and catfish and in freshwater the cichlids.

**The male Cook's soldier fish *Apogon cooki* is a mouth brooder. Note the bulge in the area below the fish's jaw which has enlarged to make room for the fertilised eggs.**

**mud crab.** Swimming crustacean and member of the family Portunidae. The mud crab *Scylla serrata,* is the largest of the 500 species of crabs found in the waters off northern Australia. Mud crabs vary in colour from green to purple depending upon the habitat, the legs are mottled and the claws have orange tips. The body and large claws yield large quantities of delicious white meat.

**The mud crab *Scylla serrata* is popular in seafood restaurants and is often held live in seawater tanks prior to sale. Note the string used to immobilise the powerful claws.**

The mud crab is most active at high tide and at night in coastal mangrove areas, feeding on hermit crabs and various molluscs such as oysters, mussels, pipis and periwinkles. It in turn is preyed upon by crocodiles, big fishes and feral pigs foraging through mangroves and mud flats. The minimum legal size is 15 cm, measured across the carapace. Large specimens are reported to reach five kilograms in weight. Common off our northern states, the mud crab can be found as far south as Sydney in varied habitats such as in mangroves, in estuaries almost up to fresh water and in the open ocean.

**The mud-oyster *Ostrea angasi* is quite edible but has a strong flavour.**

**mud-oyster.** Bivalve mollusc and member of the family Ostreidae. In the past the mud-oyster, *Ostrea angasi,* was a favourite food of

Australia's coastal Aborigines and the remains of their meals can be discovered in coastal middens. In Australia's colonial days these oysters were exploited by the convicts as food and the empty shells were burned and used as lime in the construction of buildings. The mud-oyster grows to 15 cm in length and can be found in most southern estuaries on sand and mud flats. See **giant coxcomb-oyster**, **middens**, **Pacific oyster** and **Sydney rock-oysters**.

**multi-level dive.** A dive during which bottom time is accumulated at two or more depths given in the dive tables. It is best to conduct the deepest part of the dive first and ascend to progressively shallower depths.

**mussel.** Marine bivalve mollusc and member of the family Mytilidae. Mussels are commonly eaten as food in many parts of the world. The edible mussel, *Mytilus edulis planulatus,* is common on rocky areas of coastline from New South Wales to southern Western Australia and the colour of the shell varies from brown to purple and black. The larger New Zealand green-lipped mussel, *Mytilus canaliculus,* is commonly served in restaurants. Both varieties can be purchased commercially, fresh in the shell, steamed or marinated.

**The green-lipped mussel *Mytilus canaliculus.***

**narcosis.** See **nitrogen narcosis**.

**NASDS.** See **National Association of Scuba Diving Schools Australasia Inc**.

**National Association of Scuba Diving Schools Australasia Inc. (NASDS).** The following information has been supplied by the National Association of Scuba Diving Schools Australasia Inc. Incorporating the standards and procedures of FAUI:

The National Association of Scuba Diving Schools Australasia Inc. (NASDS) was so named on the 1 July 1991 and prior to this date was known as the Federation of Australian Underwater Instructors (FAUI). The history of the formulation of the NASDS (FAUI) is closely linked with the development of the Australian Underwater Federation (AUF). Indeed the development of NASDS mirrors the evolution of the scuba diving industry in this country as a whole.

In Australia during the late 1960s the AUF emerged as the body which represented the sport of skindiving and its growing offshoot, scuba diving. At that time various clubs around the country joined to create the AUF, which controlled skindiving competitions and set sporting rules and standards, modified to meet the unique needs of the Australian scuba diver.

By the early 1970s the number of people attracted to scuba diving reached a level where specialised diving instruction on a commercial basis became economically viable. Shops specialising in the sale of diving equipment emerged, and the shop owners set up their own instructor courses based on the owner's knowledge and what was known at the time.

There were sufficient shop owners and instructors who had trained through the AUF to create an Australia-wide instructor body. Thus the NASDS (FAUI) was born and an agreement which suited both organisations was signed. NASDS agreed to adopt the AUF standards, train instructors, who in turn taught divers, while the AUF agreed to stop club based instruction. Hence the Australian diving community received professional diving instruction and this enabled national standards to be maintained, and internationally accredited certification to be issued.

NASDS grew through the 1970s, developing its technique of diver and instructor training, certification materials and texts, and register of qualified divers. An innovation produced by NASDS in the late 1970s was a package of logbooks which enabled the diver to maintain both a record of his/her diving qualifications and diving experience. Details of the NASDS course standards were laid down in these logbooks, allowing divers to know in advance exactly what was included in the course and to what level they would be taught. A refinement of the NASDS logbook system has resulted in a very comprehensive certification package.

Recent years have seen NASDS reorganise its administrative base to bring it into line with the competitive nature of the recreational scuba diving industry. NASDS maintains a large active financial membership of instructors and divemasters who elect regional and national representatives. Each of eight regions, based on the Australian States, supports one regional director who assists with administration and represents the views of the members in that region to the National Committee. The National Committee is composed of all eight regional directors plus the national director who is elected by the entire membership every three years. The National Committee is responsible for the policy making and executive functions of the NASDS and has strongly resisted attempts to amalgamate with other Australian instructor agencies, choosing to retain its democratically run non-profit association structure.

NASDS standards, have, since the association's inception, been of the highest order. NASDS has concentrated on designing progressive teaching programs and valid educational materials for Australian conditions and Australian divers. Emphasis among NASDSs instructors has always been on safety. This is demonstrated by NASDSs persistence in ignoring economic pressures, by maintaining the highest standard in instructor/student ratios, the corner-stone of safe, enjoyable, effective tuition and NASDSs record as a quality diving educator is recognised by its peers.

NASDS instructors during training must still meet the exacting requirements of the AUF, as well as the Australian Scuba Council (ASC) and the federal government through its National Coaching Accreditation Scheme. NASDS Instructors and NASDS-trained divers have also met the standards of the World Underwater Federation (CMAS). NASDS is the only instructor agency in Australia to ensure that the standards of its instructors are maintained, through an on-going re-testing programme.

Members of the Association acknowledge that they are part of a wider industry and have been active as inaugural members of the Australian Scuba Council (ASC), and the National Scuba Qualifications Commission (NSQC). NASDS has also played a positive role in the Queensland Dive and Travel Association of Australia (QDTAA), Dive Industry Travel Association of Australia (DITAA), and the Standards Association of Australia (SAA). NASDS will be moving into the 1990s with a solid base of members, professional training facilities, and products. NASDS Head Office, Unit 7, 15 Walters Drive, Osborne Park, WA 6017; or PO Box 246, Tuart Hill, WA 6060; phone: (09) 244 3500; fax: (09) 244 1462.

**National Association of Underwater Instructors (NAUI).** The following information has been supplied by NAUI AUSTRALIA:

NAUI AUSTRALIA is a non-profit educational association that, unlike many diving agencies has its board of directors elected by the members of the Association. This makes NAUI truly democratic in its structure. The purpose of NAUI is to educate the general public in the safety and techniques of participating in underwater activities. To accomplish this NAUI trains, qualifies and certifies instructors, establishes minimum standards for various programs, products and support materials to assist instructors with teaching. NAUIs primary purpose is reflected in the Association's motto 'Safety through education'.

NAUI had its beginnings in the United States of America when in 1959, the National Diving Patrol was reorganised into NAUI—a national group consisting not of divers but exclusively diving instructors. The first NAUI instructor training course was held in Houston, Texas in 1960. The first NAUI instructor training course in Australia was held at Brighton, Victoria, in 1963 and the first NAUI instructor training course conducted by Australian-based instructors was held in Brisbane in 1978.

In 1982, NAUI in Australia became NAUI AUSTRALIA, a fully operational branch. In 1986, NAUI AUSTRALIA became a separate corporate body with Peter Davidson the first full time executive director. Presently, NAUI AUSTRALIA has grown to the point of being the second largest dive training agency in the country and certainly the most prestigious.

NAUI AUSTRALIA has instant recognition throughout the world due to the affiliations with agencies, branches and chapters in the USA, West Pacific, Japan, Canada, Europe, New Zealand, Papua New Guinea and Britain.

NAUI AUSTRALIA runs instructor training courses regularly throughout the year in a variety of locations. The new successful instructors can become independent, working for themselves or working for shops and facilities all over the country. They become full voting members of the NAUI family and each controls one share of the NAUI AUSTRALIA Association.

One of the main advantages of the NAUI system is the flexibility of academic freedom. NAUI instructors are allowed and encouraged to exercise this freedom to meet the wants and needs of their students as long as the standards are met, with student safety being the prime factor.

NAUI AUSTRALIA is experiencing a tremendous growth rate and looks forward to an exciting and prolific future. Any person interested in this modern and innovative diving agency can phone or write to: NAUI AUSTRALIA, PO Box 183, Capalaba, Qld 4157; phone: (07) 390 3233, fax: (07) 390 3159; or contact: NAUI International Headquarters, PO Box 14650, Dept. DA, Montclair, CA 91763, USA; phone: (714) 621 5801.

**NAUI.** See **National Association of Underwater Instructors**.

**nausea.** A feeling of discomfort in the stomach causing a tendency to vomit.

**nautical mile.** Approximately one minute of arc of latitude and is standardised to equal 1852 metres or 6080 feet.

**nautilus.** A cephalopod mollusc of the subclass Nautiloidea with four gills, many short arms and an external spiral chambered shell. The only living members of this group which roamed the seas over 100 million years ago are three species in the genus *Nautilus* which live in the open ocean of the Indo-Pacific region, ranging from near the surface to depths in excess of 500 metres. The body of the nautilus occupies only the last chamber of the shell, gas being present in the other chambers to aid buoyancy. See **cephalopod** and **paper nautilus**.

The pearly nautilus *Nautilus pompilius.*

**neap tide.** Tides resulting from the Sun and Moon pulling at right angles, this occurs shortly after the first and third quarters of the Moon. During neap tides, low water is not as low as normal and high water is not as high as normal. In areas of large tidal variation neap tides provide the best diving conditions. See **spring tide**.

**near gale.** Winds of 28-33 knots or 52-61 km/hr.

**nematocyst.** A specialised stinging cell found only in coelenterates such as hydroids, sea-anemones, jellyfish and corals and a few animals which prey upon them. These special cells each contain a microscopic coiled thread, which is shot forward when touched, delivering a small quantity of toxin. Nematocysts are used both for collecting food and for defence. One of the most common coelenterate stingers on our shores is the bluebottle or Portuguese man-of-war *Physalia utriculus*.

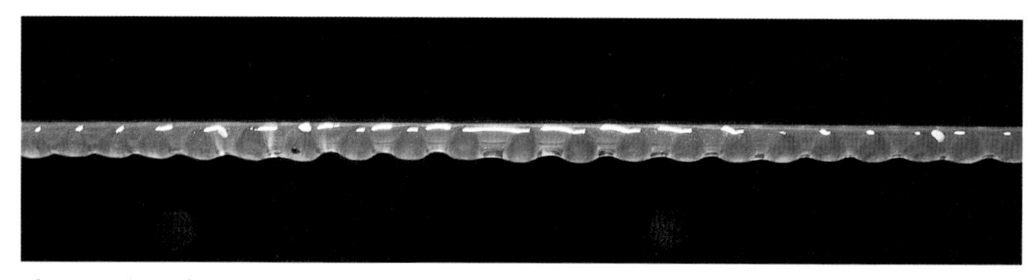

Close-up view of a Portuguese man-of-war showing an individual fishing tentacle which bears the stinging cells.

**neurotoxin.** A type of toxin which has adverse effects on nerve tissue. Saxitoxin (from shellfish) and tetrodotoxin (from pufferfish and blue-ringed octopus) are both neurotoxins. See **saxitoxin** and **pufferfish.**

**night diving.** Diving at night with the aid of an underwater torch. It is only at night that a familiar dive site can provide an exciting new diving experience. Marine creatures such as basket stars, certain fishes and various molluscs, hidden during daylight hours, emerge at night. A night dive site should be free of entanglements and fast currents, and preferably one familiar by day. Each diver should be equipped with an underwater torch and a compass. It is suggested a chemical lightstick such as a Cyalume be attached to the top of each diver's cylinder; in the event of torch failure this allows the diver to be seen underwater by the buddy and at the surface by a boat person or shore based lookout. A light and observer should be posted on shore at the entry point to enable easy exit from the water. Night diving courses are organised by many dive shops, dive clubs and individual instructors, to teach the safest techniques. See **Cyalume**.

**Queensland photographer Bill Wood night diving in the Coral Sea with an olive sea-snake.**

**Nikonos camera.** A self-contained underwater 35 mm camera. Nikon, the manufacturer of the Nikonos camera, has released a new model designated the Nikonos RS-AF which is the world's first self-contained autofocusing underwater SLR. This revolutionary new camera allows viewing and focusing of the subject through the lens via the high eyepoint action viewfinder. Other features include: freeze focus, power manual focus, matrix metering, integral motordrive and focal plane shutter with speeds from one to 1/2000 of a second (plus B setting). A selection of three lenses presently available include a 20–35 mm f2.8 zoom, a 28 mm f2.8 wide angle and a 50 mm f2.8 Micro-Nikkor. The SB-104 speedlight (underwater electronic flash unit) completes the outfit with features such as through the lens (TTL) centre-weighted flash, matrix balanced fill flash, and rear-curtain flash synchronisation. This speedlight has an underwater Guide Number of 16, in metres at ISO 100. Please note that the new Nikonos RS-AF will not replace the Nikonos V but will offer an alternative for the more advanced photographer.

The latest model Nikonos is designated the Nikonos RS-AF and was released in mid 1992.

The SB-103 strobe.

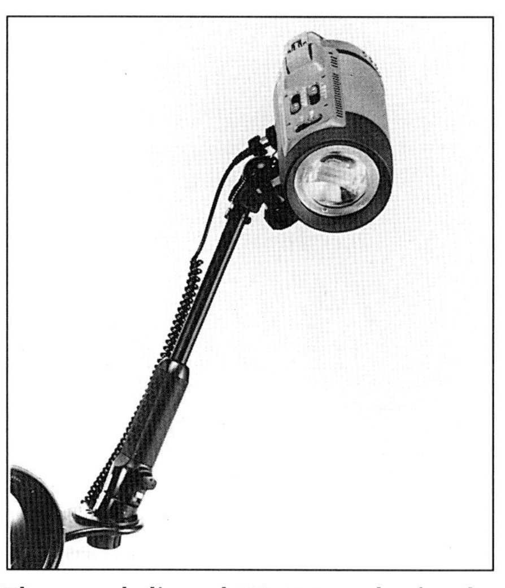
The new dedicated SB-104 strobe for the Nikonos RS-AF.

The first Nikonos camera was almost a direct copy of the Calypso-Phot underwater 35 mm camera. The patent owner Jacques-Yves Cousteau approached the Nikon company in Japan to manufacture a variation of the Calypso-Phot and this resulted in the first Nikonos camera in 1963. Cousteau held the patent for this type of camera until 1988 when it expired. The history of the Nikonos range of cameras is as follows: the first Nikonos 1963, Nikonos II 1968, Nikonos III 1975, Nikonos IVA 1980, Nikonos V 1984, and the Nikonos RS-AF 1992.

Nikonos II, 1968.

Nikonos III, 1975.

Nikonos IVA, 1980.

Nikonos V, 1984.

The Nikonos V has a 35mm lens as standard and offers automatic or manual exposure modes with exposure information displayed in the viewfinder. There is a large range of accessories available (for the Nikonos V and earlier models) including: Nikkor 80 mm f/4 telephoto lens, UW-Nikkor 15 mm f/2.8, UW-Nikkor 20 mm f/2.8, UW-Nikkor 28 mm f/3.5, LW-Nikkor 28 mm f/2.8 wide angle lenses, a slip-on Nikonos close-up lens outfit with three different field framers for various lenses, and two Speedlights: the SB-102, the more powerful of the two, with a built-in modelling light and; the SB-103, specially designed for close-up photography. The W-Nikkor 35 mm f/2.5 standard lens and the Nikkor 80 mm telephoto lens can also be used above water. The specially designed LW-Nikkor 28 mm f/2.8 is not for underwater use but is watertight and has been designed for surfing and marine sports photography. The distributor for the Nikonos system in Australia is: Maxwell Optical Industries, 100 Harris St, Pyrmont, NSW 2099; phone: (02) 660 7088, fax: (02) 660 8739. See **Calypso-Phot**.

**Ningaloo Marine Park.** A marine national park in Western Australia. This park was declared in 1978 as a joint project of the Commonwealth and Western Australian governments. The coral reef extends 260 km between North West Cape and Amherst Point and covers 4300 square kilometres. Over 200 species of coral, 460 species of fishes and 600 species of molluscs have been identified in the area. Ningaloo has Australia's largest fringing reef and owes its existence to the absence of large rivers on the Western Australian coast. Ningaloo lies in the route of migrating humpback whales, which spend the summer months in the Antarctic regions feeding on large schools of krill and other plankton. For information on the Ningaloo Marine Park, contact the Department of Conservation and Land Management; phone: (09) 367 0333.

**Ningaloo Marine Park (Pilgrammuna Bay) is ideal for small trailer-operated dive boats as the outer reef edge is relatively close to shore.**

**nitrogen.** A colourless, odourless, tasteless gas constituting 79% of the atmosphere by volume. When scuba divers are at depth, the nitrogen they take in with each breath of air can have a narcotic effect similar to laughing gas (nitrous oxide) only much weaker. This phenomenon is called nitrogen narcosis. See **nitrogen narcosis**.

**nitrogen narcosis.** The narcotic effect produced when air is breathed at depth due to the high partial pressure of nitrogen. This effect has been given a number of names such as: 'raptures of the deep', 'the narks' and 'depth intoxication'. Narcosis due to nitrogen places a limitation on safe diving with ordinary compressed air to approximately 40–50 metres. Deeper diving requires the substitution of a inert gas such as helium. Symptoms of nitrogen narcosis are: a sense of well-being, no concern for personal safety, giddiness and lack of coordination. The effects may be similar to those of alcoholic intoxication. Nitrogen narcosis has often caused diving fatalities and symptoms can occur in as little as 20 metres of water. Divers suffering from nitrogen narcosis do irrational things such as: offering their regulator to a passing fish, ignoring buddy's signals and overstaying predetermined bottom times. Narcosis is more likely if the diver is suffering a hangover, or is diving in cold water, or is taking prescription drugs or other medications. Nitrogen narcosis most commonly affects those unused to diving at depth; as with alcohol in social situations, exposure, and incremental increases to depth, produces a built-up resilience to its effects. See **nitrogen**.

***Noctiluca.*** A spherical transparent pink, bioluminescent single-celled marine dinoflagellate about 1mm in diameter. Member of the order Dinoflagellata, genus *Noctiluca.* This dinoflagellate produces luciferin, a biochemical substance responsible for light production in the sea. *Noctiluca* is found worldwide. See **bioluminescence**.

**no-decompression dive.** A dive that does not exceed the no-decompression limits of your dive table. A no-decompression dive is one where the diver's bottom time is such that no decompression stop is required before surfacing. A safe habit for sport divers is to make all dives a 'no-decompression dive'. A safety stop at three to four metres for a few minutes is a good idea on every dive, no matter how shallow the dive depth.

**no-decompression limit.** The maximum allowable bottom time that a diver may spend at a given depth without having to conduct a decompression stop before surfacing.

**nori.** Edible red alga and member of the genus *Porphyra* which grows in delicate membraneous sheets. Nori is cultivated in Japan where many thousands of tonnes (wet weight) are collected annually and dried on bamboo mats. The processed alga has the appearance of thick sheets of purple-brown paper and is sold in this form, to be used in soups and in rice dishes. Some species of *Porphyra* can also be found in Australian waters.

**nudibranch.** A shell-less mollusc and member of the subclass Opisthobranchia, order Nudibranchia. Nudibranchs are usually characterised by external and naked gills on the dorsal (back) surface and are some of the most colourful underwater creatures. Nudibranchs are found in all oceans of the world including under the Antarctic ice. There are approximately 3000 species worldwide and probably up to 800 species in the Indo-Pacific area. Individuals have a short life span, ranging from as little as a month to as much as a year and, being hermaphrodites, are able to mate with any mature individual of the same species. *Nudibranchs of Australasia* by Richard C. Willan and Neville Coleman; published by Australasian Marine Photgraphic Index, Sydney 1984 is a good reference text with hundreds of colour photographs to make identification easier. See **hermaphrodite** and ***Glacus atlanticus***.

**The nudibranch *Ceratosma cornigerum* belongs to the family Chromodorididae and varies in colour depending upon the location.**

**The nudibranch *Polycera capensis* feeds mainly on the blue-green coloured bryozoan *Bugula dentata* and is common in New South Wales coastal waters from Broken Bay north to Port Stephens.**

**numbfish.** A relative of the skates and stingrays and member of the family Torpedinidae. The numbfish has specialised electric organs running along its dorsal surface, starting almost between the eyes and extending back past the last gill-slit. The electric ray *Hypnos monopterygium,* commonly known as the numbfish, is found on sandy and muddy bottoms in most Australian States with the exception of the Northern Territory. The electric organs consist of up to 500 hexagonal cells arranged in a honeycomb pattern, each cell containing a

The short-tailed electric ray *Hypnos monoterygium.*

jelly-like substance which acts as a storage battery, with the positive pole on the upper surface and the negative pole on the ventral or lower surface. These cells can deliver up to 200 volts, though this drops progressively to about 80 volts with successive rapid discharges. For more information on this fish, see Fishwatcher's Notebook in the February/March 1986 issue of *Scuba Diver* magazine. See **marine leech**.

**ocean.** The body of salt water which covers three quarters of the earth's surface.

**oceanarium.** A salt-water pool or enclosed part of the sea in which marine creatures such as dolphins and small whales are kept.

**Oceania.** A collective name used to describe all the Pacific islands. Australia is sometimes included.

**oceanographer.** A person who studies physical, geographical, chemical or biological aspects of the ocean.

**octocoral.** A soft coral and member of the subclass Alcyonaria or Octocorallia. Octocorals are bottom-living and colonial and they include sea-pens, sea-fans, sea-whips and many others. Individual octocoral polyps have eight feathery tentacles which connect to the stomach cavity. Octocorals live in colonies supported by internal calcareous spicules or a proteinaceous axis. See **soft coral**.

**octopus.** A cephalopod mollusc and member of the order Octopoda. Octopuses have a short rounded body and a head with eight tentacles joined at the base by a web of skin. They are commonly found on rocky and sandy bottoms hiding in holes and crevices by day, and feeding on molluscs, crustaceans and fishes at night. They have the ability to expand and contract special pigment cells in their skin to affect colour changes to match their surroundings. An intelligent species they quickly learn 'tricks' such as in unscrewing a cap from a bottle to reach food placed inside.

**A pair of octopuses, *Octopus cyaneus*, outside their rocky lair.**

**octopush.** A form of hockey played underwater in a swimming pool by players equipped with mask, snorkel and fins. A 'pusher' is used to move the lead 'squid' for scoring points called 'gulls'.

**octopus regulator.** A spare second stage regulator mouthpiece attached to the first stage reducing valve for use in an emergency. Buddy breathing is difficult to carry out in rescue situations and the octopus system was introduced to overcome this and other problems such as a malfunction in the second stage of a regulator. This system enables a second diver to breathe from the same scuba cylinder without the need to share the primary regulator.

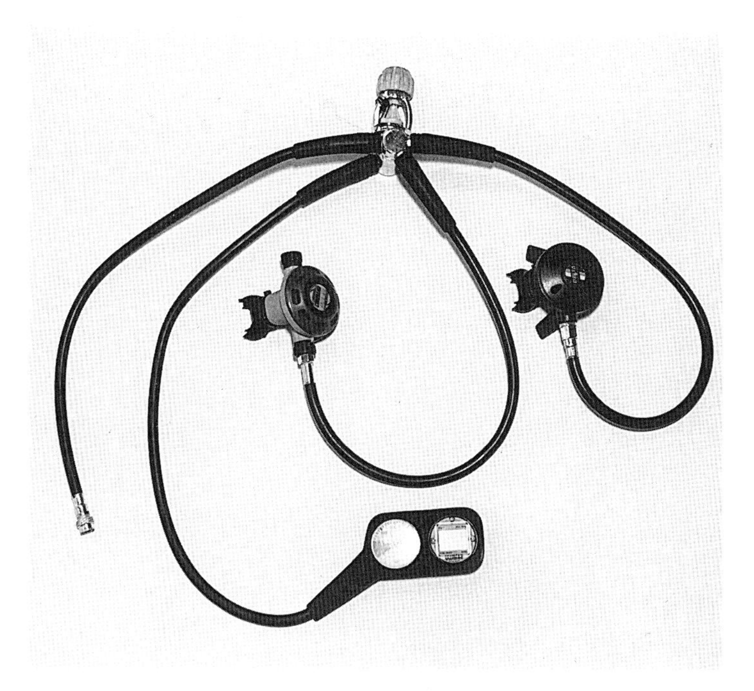

An octopus regulator.

**omnivorous.** Eating a diet of both plants and animals.

**open circuit.** A type of regulator which exhausts the air to the surrounding water after a single breath rather than back into the system. Scuba demand valves fall into this category.

**open water.** A river, sea or lake. Also a term used to indicate diving in naturally occurring water as opposed to a swimming pool.

**operculum.** A shield of a horny or calcareous substance which is used to block the entrance in the shell of a gastropod mollusc. As the animal withdraws into its shell, the operculum closes over the opening and forms a barrier between it and the environment, providing protection from predators and conserving moisture if exposed to air at low tide.

**opisthobranch.** A gastropod mollusc and member of the subclass Opisthobranchia. Included in this group are nudibranchs, sea-hares and bubble shells. Nearly all opisthobranchs are hermaphroditic, and lay eggs in jelly-like masses which hatch to form free-swimming larvae. See **hermaphrodite**, **nudibranch** and **sea-hare**.

**orca.** A whale and member of the order Cetacea, suborder Odontoceti (toothed whales), family Globicephalidae. Of the toothed whales the orca *Orcinus orca* is second in size only to the sperm whale. Commonly called 'killer whales', orcas live in all oceans of the world. They are superpredators, eating fishes, squid, sharks, dolphins, seals, walrus, sea-lions and sea birds including penguins. Their colour is a striking patchwork of black on the dorsal surface and white on the underside, with an oval white patch on both sides near the head. In a large pack called a 'pod' they attack and kill other whales, eating only the lips and tongue of the larger species. Male orcas can grow to nine metres in length and weigh up to seven tonnes. The erect dorsal fin can reach nearly two metres in height. The female of the species is smaller and is distinguished by a shorter hooked dorsal fin. In the 19th century three pods of orcas combined with shore-based whalers to kill larger species of whales at Eden on the south coast of New South Wales—one of several recorded instances of such co-operative hunting alliances.

**Organisation for the Rescue and Research of Cetaceans in Australia (ORRCA).** The following information has been supplied by ORRCA:

The Organisation for the Rescue and Research of Cetaceans in Australia (ORRCA) was established in 1985 to assist the National Parks and Wildlife Service at whale and dolphin strandings, it operates on a volunteer non-profit basis. ORRCA is also involved in the rescue of other marine mammals such as seals and sea-lions.

Rescue Training Workshops are regularly attended by ORRCA members and by officers of the National Parks and Wildlife Service. There is a variety of other activities, ranging from whalewatch cruises to a public awareness campaign about the fact that large numbers of whales are dying of starvation after ingesting plastic bags in the sea. Members keep abreast of developments in ORRCA and the wider cetacean world through a monthly newsletter.

Little is known of whale and dolphin behaviour in the wild, and ORRCA is engaged in learning more about this fascinating subject. The bottlenose dolphins of Jervis Bay, NSW, are the focus of an ORRCA research project. Whales and dolphins frequently strand on our coastline, sometimes more than a hundred at a time. There is now ample evidence that mass strandings result from social bonds between a stranded individual and the rest of the group, and are not 'suicides' as was once believed.

In recent years, it has been found that these animals can be saved if the correct techniques are used. For example, of the 114 false killer whales stranded at Augusta, Western Australia, 96 were returned to the sea in good condition. Australia is recognised as the world leader in the art of whale rescue. Without knowledge of the correct techniques, rescue attempts unfortunately do more harm than good in most cases. ORRCA, PO Box E293, St. James, NSW 2000; phone: (02) 667 0048. To report whale strandings on the 24 hour hotline phone: (02) 412 4747.

**organ pipe coral.** A coral and member of the subclass Alcyonaria or Octocorallia, order Stolonifera, one of only two soft corals in Great Barrier Reef waters that secrete hard limestone skeletons. The organ pipe coral *Tubipora musica,* forms a series of parallel long thin red calcareous tubes connected by horizontal platforms, the whole structure resembling organ pipes. The polyps are grey-green and are extended both day and night. As the coral skeleton retains its colour even after death (one of the few corals to do so) it is often sold as an ornament and is used in making jewellery.

**O-ring.** A sealing ring generally of rubber or synthetic material, commonly used in the construction of underwater equipment such as camera housings, regulators, valves and compressors to seal and keep water and other fluids out of various parts of the equipment.

**ORRCA.** See **Organisation for the Rescue and Research of Cetaceans in Australia**.

**otolith.** Earstones of fish, formed from the layered deposition of calcium carbonate. See **age: determination of fishes**.

**oviparous.** An egg-laying animal which hatches its eggs outside its own body.

**ovoviviparous.** An animal which hatches its eggs inside its own body. The young are born alive but without placental attachment.

**oxygen.** A colourless, odourless and tasteless gas comprising approximately 21% of the earth's atmosphere.

**oxygen poisoning.** See **oxygen toxicity**.

**oxygen rebreathing set.** See **closed circuit breathing apparatus**.

**oxygen toxicity.** Elevated oxygen pressure can cause oxygen toxicity in most individuals when the partial pressure of oxygen breathed exceeds two atmospheres absolute. This can cause a loss of consciousness and may result in drowning. It can occur in the following situations: breathing 100% oxygen at depths in excess of 10 metres, and breathing compressed air at depths in excess of 90 metres. For depths greater than 70 metres, commercial diving operations progressively reduce the percentage of oxygen in the heliox mix to eliminate this problem.

**oyster.** See **Sydney rock-oyster**, **mud-oyster**, **Pacific oyster**, **giant coxcomb-oyster** and **middens**.

**Pacific oyster.** A bivalve mollusc and member of the family Ostreidae. The Pacific oyster *Crassostrea gigas* is cultivated widely around the world and is also known as the Japanese oyster because it is native to Japan. The Pacific oyster was introduced into Tasmanian waters in 1947 and has flourished there ever since. New South Wales oyster-farmers fear that this fast growing oyster which can reach a length of 15 cm in four years will take over from the much slower growing but highly prized Sydney rock-oyster.

**PADI.** See **Professional Association of Diving Instructors (PADI)**.

**painted crayfish.** See **tropical rock lobster**.

**panic.** Excessive anxiety, caused by one or a combination of factors, such as seasickness, fatigue, fast currents, deep diving and equipment problems such as a leaking face mask or hard breathing regulator. If you feel you are anxious about your present situation then you should STOP whatever you're doing, and SLOW your breathing, so it is longer and deeper. After doing this you may be able to control your panic and reassess the situation. Your options may then include terminating your dive and making your way to the surface or continuing the dive.

**paper nautilus.** Cephalopod mollusc and member of the subclass Coleoidea, order Octopoda. The paper nautilus *Argonauta nodosa,* is also called the argonaut shell and is common in Bass Strait and, during the spawning season, off

**The paper nautilus *Argonauta nodosa* secretes a fragile shell which is popular with collectors.**

Montague Island in southern New South Wales. It is octopus-like in shape with eight arms and the female secretes a fragile white shell by means of specialised glandular arms, the shell serves as protection and for brooding the eggs. The male of the species does not secrete a shell. See **cephalopod** and **nautilus**.

**paralytic shellfish poisoning (PSP).** See **saxitoxin**.

**partial pressure.** The pressure exerted by one gas only in a mixture of gases. If the ambient pressure of air is 1.0 atmosphere absolute, the partial pressure of oxygen in dry air is 0.2099 atmospheres or expressed as a percentage, 21%. Therefore if air contains 21% oxygen, the oxygen exerts 21% of the total pressure of air. See **Dalton's law**.

**pearl.** A hard, smooth, lustrous concretion secreted within the shell of various bivalve molluscs, usually round or oval in shape and often valuable as a gem. Formed when small foreign bodies are introduced either naturally or artificially into bivalves such as the black-lipped pearl shell *Pinctada margaritifera* and the gold-lipped pearl shell *Pinctada maxima*. Pearls are formed over a number of years by the shell's mantle which, in response to irritation by the foreign body, secretes concentric glossy layers of mother-of-pearl around the object to form a pearl. Apart from white pearls there are also pink, black and bluish varieties. Freshwater mussels are also seeded to produce 'freshwater pearls'. Experiments with seeding abalone have also produced pearls of good quality.

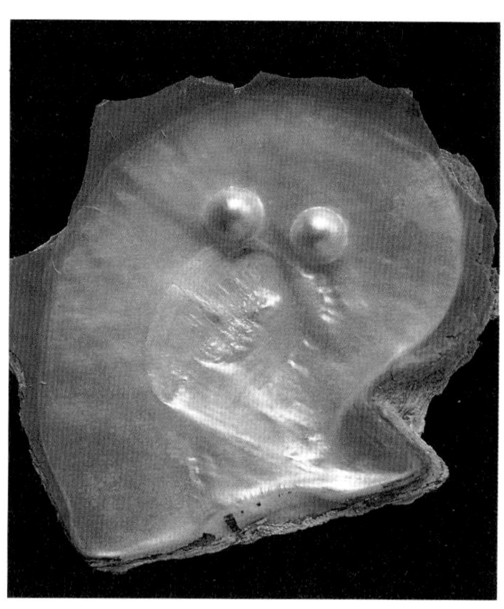

**The gold-lipped pearl shell *Pinctada maxima* showing two cultured half pearls.**

**pectoral.** Pertaining to the breast or chest. In fishes a pair of fins found behind the head, one on each side.

**pedicellariae.** Small jointed spines on the surface of sea-stars and sea-urchins which are arranged in sets of two and three and act as pincers. The pedicellariae of certain species of sea-urchins can inject venom, causing severe pain. See **sea-urchin**.

**pelagic.** Marine plants and animals living at or near the surface of the ocean as opposed to those living on the bottom.

**pelagic nudibranch.** See ***Glaucus atlanticus***.

**Perciformes.** Perch-like fishes of the order Perciformes. This order of bony fishes has the largest number of species and is divided into 20 sub-orders.

**periostracum.** The epidermis or outer skin of some molluscs. Shell collectors can remove this layer by soaking shells in household bleach so the true colours and brilliance of the shell's undersurface can be revealed.

**personal dive sonar.** An underwater hand-held sonar device used for checking depths and distances to objects underwater. Scubapro's dive sonar allows divers to check the depth of the water before entering by simply holding the sonar (which is about the same size as a small torch) under the surface for a few seconds. The sonar beam of 24 degrees allows objects to be pinpointed up to 79 metres away and operates down to depths of 91 metres.

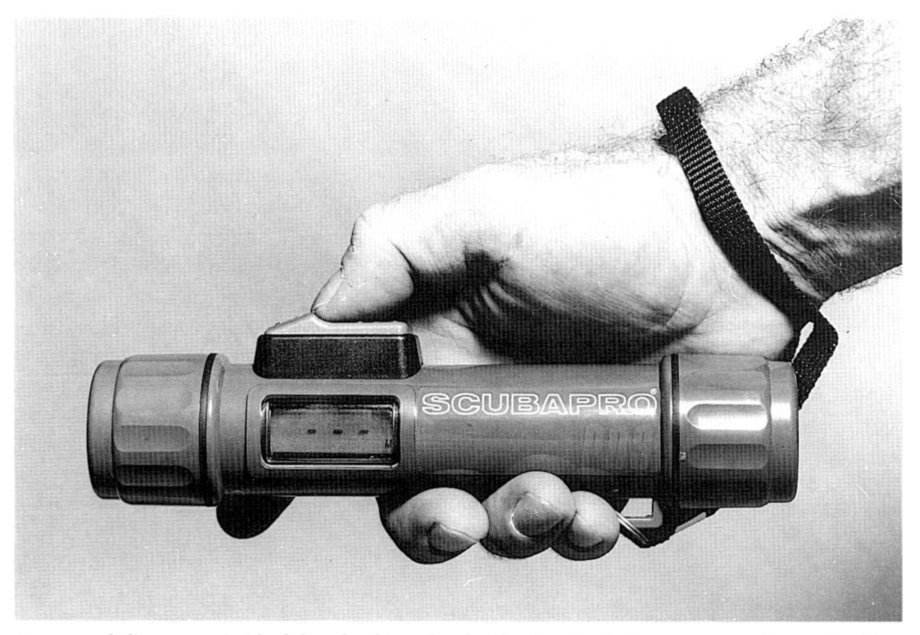


**A personal dive sonar is ideal for checking the depth of water before commencing scuba dives from small boats which lack depth sounders.**

**phycologist.** A person who studies algae or seaweeds.

**phylogeny.** The evolutionary relationship and ancestry of living organisms.

**phylum.** One of the principal divisions employed within the phylogenetic classification of the Animal Kingdom. See **scientific classification of marine organisms**.

***Physalia utriculus.*** See **Portuguese man-of-war**.

**phytoplankton.** Collective name for small single and multi-celled plants such as diatoms and dinoflagellates which live in the ocean or lakes. These tiny plants produce organic material by means of photosynthesis and are more common at or near the surface. All species of marine animals are reliant on phytoplankton either directly or indirectly for food. See **plankton** and **zooplankton**.

**Piccard, Jacques.** Swiss-born oceanographer who made the deepest manned ocean descent to a depth of 10916 metres, in the bathyscaphe *Trieste* in the Marianas Trench 400 kilometres southwest of Guam on 23 January 1960, with Lt. Donald Walsh of the United States Navy.

**pillar valve.** The tap or valve which is necessary to turn high pressure cylinders, such as scuba cylinders, on and off. Also called a K-valve. See **J-valve** and **K-valve**.

**pipi.** A bivalve mollusc and member of the family Donacidae which live intertidally only a few centimetres below the surface of shallow sloping sandy ocean beaches. The pipi *Donax deltoides* is a common species in New South Wales waters and the shell has a smooth polished outer surface which may be coloured in various shades of pink, olive green, yellow or lavender. It can grow up to eight centimetres in length and is found from southern Queensland along the eastern and south-eastern coast of Australia, around Tasmania, and the coast of South Australia to southern Western Australia. Fishermen collect pipis for bait and they are sometimes served in seafood stews and soups. Pipi shells can be found in middens along Australia's coastline, an indication that this shellfish formed a regular part of the diet of the early Aborigines. See **midden**.

**The pipi *Donax deltoides* in the process of burrowing into wet sand.**

**pitchpolled.** A nautical term used to describe a ship which has been somersaulted stern over stem or vice versa by a heavy sea. See **stem** and **stern**.

**plankton.** A collective name for plants and animals that drift with the ocean currents. Planktonic organisms include single-celled and multi-celled algae, bacteria, invertebrates, the larvae of vertebrates and eggs. Animal plankton is called zooplankton and plant plankton is termed phytoplankton. The largest of all animals, the blue whale, feeds exclusively on plankton and so does the largest species of shark—the basking shark. Both have mouths which are designed to sieve plankton from the sea. See **phytoplankton**, **zooplankton**, **whale** and **whale shark**.

**plankton bloom.** See **red tide**.

**poikilothermic.** A term describing all animals whose body temperature changes with the environmental temperature. These animals are commonly called cold-blooded e.g. fishes.

**polychaetes.** Bristleworms, fan-worms and tube-worms of the class Polychaeta. Polychaetes are annelid worms and among the most common animals of seashores but they usually remain hidden under rocks, in constructed tubes and in burrows in the sediment; so they are not evident to the casual observer. In tropical waters the Christmas tree worm is notable for the variation in its bright colouration and the speed at which it withdraws into its tube when disturbed.

**Christmas tree worms *Spirobranchus giganteus* build calcareous tubes deep within coral heads and withdraw into them in times of danger.**

**polyp.** An individual zooid of a compound or colonial organism characterised by a columnar body, and a dorsal end containing a mouth and tentacles. Coelenterates such as corals can have a single polyp or a number of polyps.

**Soft coral polyps have eight feathery tentacles.**

**porcupinefish.** A toxic species of fish and member of the family Diodontidae. Porcupinefish have the ability to inflate their bodies by swallowing water (or air if out of water) like their close relations the pufferfish (family Tetraodontidae). Porcupinefish however, differ from pufferfish by the presence of stiff spines embedded in the skin. In addition, the teeth are fused into an undivided beak, whereas the fused teeth of the pufferfish are split at the front of the jaws. Porcupinefish should never be eaten as the flesh is poisonous. See **pufferfish**.

**The freckled porcupine fish *Diodon holocanthus*.**

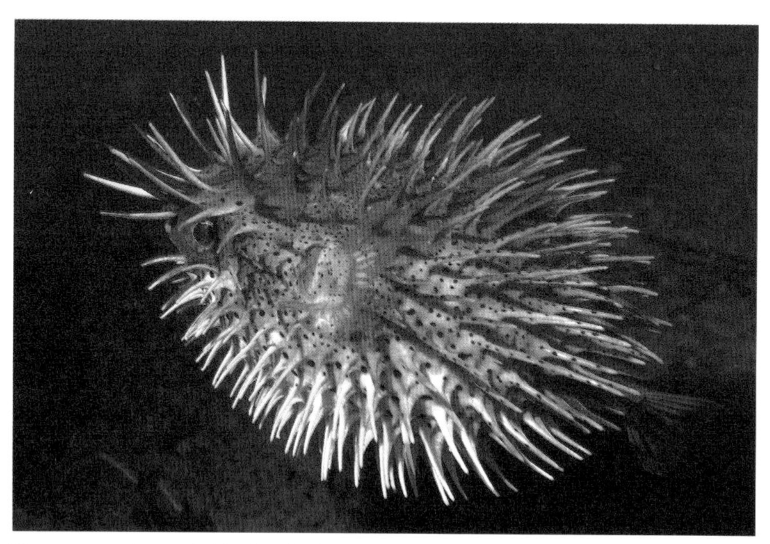

**The erect spines of this freckled porcupine fish *Diodon holocanthus* act as a deterrent to most predators.**

**Porifera.** See **sponge**.

***Porpita porpita.*** Coelenterate and member of the class Hydrozoa, suborder Chondrophora. This floating oceanic coelenterate is comprised of a flat circular disc about two centimetres in diameter with a central feeding zooid or mouth, fringed on the extreme edges by short defensive tentacles. *Porpita porpita* is actually a colony of animals which gives the appearance of a single animal. It is often found washed up on beaches in association with the hydrozoan *Velella vella* and the predatory molluscs *Glaucus atlanticus* and *Janthina* spp. See **by-the-wind sailor**, ***Glaucus atlanticus*** and ***Janthina* spp**.

**Port Jackson shark.** Member of the order Heterodontiformes, family Heterodontidae. Eight species are known worldwide and there are two common species in Australian waters; they are the common Port Jackson shark *Heterodontus portusjacksoni* which is found in southern coastal waters and *Heterodontus galeatus* the crested Port Jackson shark which is less common and is found from southern Queensland to Batemans Bay in New South Wales. Egg cases of Port Jackson sharks can be hatched in the home aquarium. For more details see Fishwatcher's Notebook in the June/July 1987 issue of *Scuba Diver* magazine.

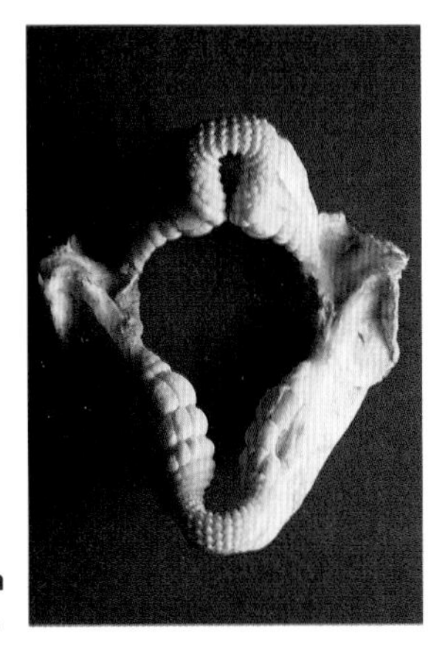

**The jaws of the Port Jackson shark are filled with plate-like teeth which are used for crushing their prey, mainly molluscs and crustaceans.**

**Schools of Port Jackson sharks *Heterodontus portusjacksoni* congregate at certain times of the year to mate.**

The crested Port Jackson shark *Heterodontus galeatus.*

**Portuguese man-of-war.** A stinging coelenterate and member of the class Hydrozoa, order Siphonophora. The Portuguese man-of-war or *Physalia utriculus* is commonly called a bluebottle and is widely distributed in tropical and temperate seas and floats on the surface with the aid of a gas-filled float. It is commonly seen when the wind is blowing onshore, and may be encountered by scuba divers or snorkellers at sea or by swimmers when it is washed onto beaches. This blue coloured stinging hydrozoan has a double-ended float up to 12 cm long and long trailing tentacles that may reach 12 metres in length. It is well armed with stinging cells, or nematocysts, which if touched can cause intense pain and red welts that may last for weeks. Vinegar is the best first aid treatment. Bluebottles feed on small fishes and zooplankton which are stung and paralysed with a potent neurotoxin from the nematocysts on the trailing tentacles. The tentacles then retract to bring the food within reach of the feeding polyps beneath the float for

The Portuguese man-of-war *Physalia utriculus.*

digestion. Predators of the bluebottle include: *Glaucus atlanticus* (pelagic nudibranch), turtles, crabs, *Janthina* spp. (violet sea snail) and the ocean sunfish. Bluebottles are often found washed up on beaches in association with *Glaucus atlanticus, Velella velella,* (by-the-wind sailor) and *Porpita porpita* (floating oceanic coelenterate). See **by-the-wind sailor**, ***Glaucus atlanticus***, ***Janthina* spp.** and ***Porpita porpita***.

***Posidonia australis.*** See **sea-grass**.

**powerhead.** An explosive device used for killing sharks and large fish, consisting of a short metal tube, firing pin, spring and ammunition, which is attached to a short stick, speargun or handspear. Powerheads are manufactured to suit a variety of ammunition including: 12 gauge shotgun cartridges, .303 and 30-30 calibre rifle cartridges. The fish is killed by the explosion and shock wave when the tip of the powerhead physically contacts its body, causing the cartridge to be forced back onto the firing pin and discharged. The powerhead was popular during the 1960s-1970s and was used to slaughter large numbers of grey nurse and other sharks. Also called a bangstick. See **shark billy**.

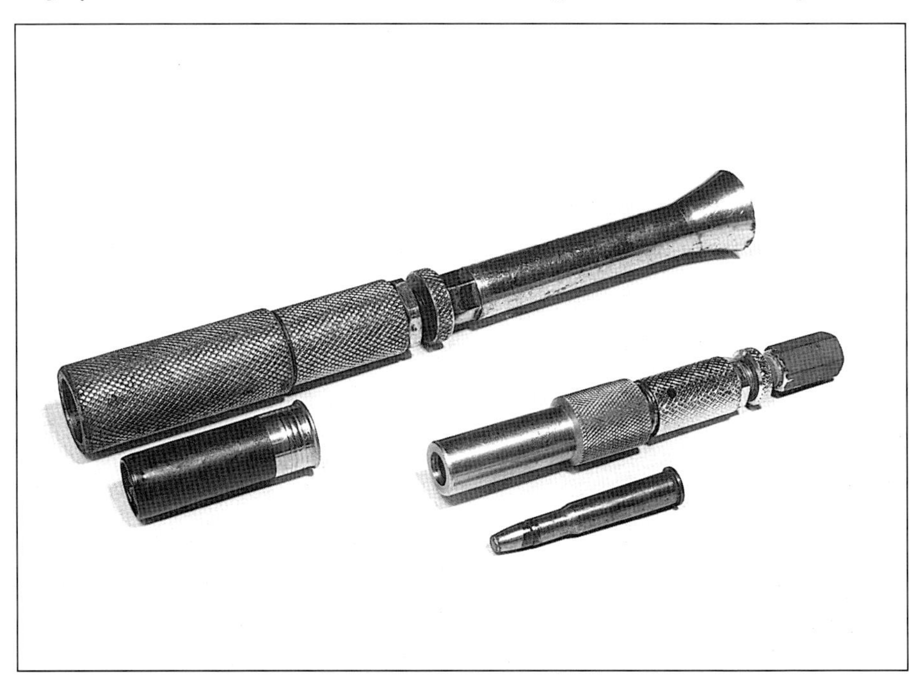

**A range of powerheads which can be attached to handspears or spearguns and used for killing large fish and sharks.**

**pregnancy and diving.** The effects of pressure on the foetus are largely unknown and unresearched, so diving when pregnant is not advisable, especially if taking drugs for nausea.

**pressure conversion table.**

1 atmosphere = 14.7 pounds/sq. inch (psi)
= 1.033 kilograms/sq. cm ($kg/cm^2$)
= 101.325 kilopascals (kPa)
= 1.013 bars
= 10.07 metres sea water
= 33.05 feet sea water
= 760 millimetres mercury (mm Hg)

**proboscis.** Any elongate, snoutlike tubular feeding organ.

**Professional Association of Diving Instructors (PADI).** The following information has been supplied by PADI Australia:

The Professional Association of Diving Instructors (PADI), is the world's largest and fastest-growing diving organisation. With over 20000 active instructors worldwide, who certify over 320000 students a year and with more than 1500 training facilities, PADIs position as the leader in diver education has tremendous impact on the growth and direction of the diving industry. With a headquarters staff of over 150 full-time professionals, branch offices in Australia, Canada, Switzerland, Japan, New Zealand, Norway and Sweden, and courses conducted in over 80 countries, PADI is truly an international organisation.

Established in 1966, PADI was formed under the leadership of Ralph Erickson and John Cronin, highly respected experts in the field of diver education. Since then PADI has achieved exceptional growth by providing quality education to an ever-expanding number of diving consumers and by helping others succeed as diving professionals.

PADI began by recognising the need for more effective instructor training. Traditional instructor certification had been based on testing knowledge, diving ability and physical fitness endurance. While PADI believed these factors were not unimportant, its alternative approach stressed training in *how to teach* in the pool, classroom and open-water environments.

PADI instructors are developed in two phases—the Instructor Development Course and the Instructor Evaluation Course. The training phase is comprised of 26 modules and is conducted by some of the most highly skilled educators in the industry—PADI Course Directors. The second phase is purely evaluative and is conducted by qualified Instructor Examiners from PADI headquarters. The rigorous training and evaluation results in the most sought-after diving instructors in the world.

To keep members up to date, PADI headquarters staff regularly conduct a variety of instructor-development seminars at locations throughout the world. These programmes cover topics that range from new methods for conducting entry-level diver training to how to promote and conduct continuing education courses.

Within two years of its birth, PADI had created one of the many products that would have major impact on the entire diving industry. The PADI Positive Identification Card (PIC), introduced in 1968, was the first identification credential to include the diver's picture and personal identification information.

PADIs continuing education courses are based on the concept that divers cannot be taught everything in a single course. Continuing education provides a step-to-step series of courses that result in competent, safe and enthusiastic divers. The steps in the PADI system include Open Water Diver, Advanced Open Water Diver, Rescue Diver, Divemaster, Specialty Diver and Master Scuba Diver ratings.

The heart of the PADI system is the Open Water Diver course. Established in 1972 as the preferred entry-level training course, this highly acclaimed course places emphasis on training and experience in open water. The Open Water Diver course includes four open water scuba dives in addition to theory and pool work.

In 1977, the Open Water Diver programme was further enhanced by the creation of the PADI Modular Scuba Course. This totally coordinated program, consisting of an instructor manual, student text, recreational dive planner, standardised quizzes and exams, an audiovisual and aquatic cue cards made teaching a consistent, high-quality diver training programme easy. Today, the PADI Modular Scuba Course components are the most widely used diver-training materials in the world.

Over the years, PADI has looked for new ways to make learning to dive easier, more enjoyable and more accessible. The Discover Scuba experience, which debuted in early 1986, is the result of that research. The first and only of its kind in the industry, Discover Scuba is designed to allow people to try scuba—without the cost and time involvement of an entire course, and in the comfortable surroundings of a local swimming pool.

PADI continues to strive for new and better ways to improve diving—both as a recreational activity for consumers, and as a rewarding career for professionals. This commitment to help make diving a strong and dynamic industry is what has lead to PADIs past, present and future growth. PADI Australia, Unit 1 No 1-7 Lyon Park Road, North Ryde, NSW 2113; phone: (02) 888 5899, fax: (02) 805 0870, or: PADI International Headquarters, 1251 East Dyer Road #100 Santa Ana, CA 92705-5605, USA; phone: (714) 540-7234, fax: (714) 540 2609.

**Project Jonah.** A conservation project concerned with stopping commercial whaling operations all around the world. Contact: 340 Gore Street, Fitzroy, Vic. 3065; phone: (03) 416 1455, fax: (03) 416 0767, toll free number phone: (008) 332 510.

**Project Stickybeak.** A confidential, non-punitive, medically controlled scheme that collects information on all types and severities of diving-related problems.

The majority of reports so far have been concerned with fatalities. However the intent is to obtain reports of minor problems also, in order to identify potentially dangerous situations before fatalities occur. Provisional reports are prepared and are made available to the major diving organisations. Reports are also printed in the Journal of the South Pacific Underwater Medical Society (SPUMS) which has an international distribution. Published reports do not contain details which identify those concerned in the incident.

This project, which has been operating for more than 15 years, has the support of all the main diving organisations. The success of the scheme, however, depends on the supply of reports by divers and is therefore dependent on their appreciation of the value of reports in indicating problems. It is believed that the provisional reports on fatalities and the occasional reports of non-fatal incidents has been shown to be beneficial.

The scheme is now set to obtain reports from hyperbaric units in Australia, New Zealand and overseas and is seeking to include other data from overseas on all types of diving problems in a database file that will be accessible to all persons and organisations with an interest in diver safety. This database will utilise both IBM and Apple Macintosh systems and the data will be entered from reports which have first had the identifying details removed. Reports and correspondence should be directed to: Dr D. Walker, PO Box 120, Narrabeen, NSW 2101.

**propeller guard.** A special cage which encloses the propeller of surf rescue boats, preventing injury to people being rescued by power boats in the surf. This was developed for the Royal Life Saving Association of Australia.

**psi.** An acronym for pounds per square inch (pressure). See **pressure conversion table**.

**public marine aquaria.** See **marine aquaria public**.

**pufferfish.** A toxic species of fish and member of the family Tetraodontidae. All the fishes in this family are without true scales and are also called toadfishes. When pufferfish are handled or frightened they gulp water (or air if out of water), this causes their highly elastic skin to stretch and the body to swell—hence the common name of puffer. Pufferfish should never be eaten as the flesh, especially the viscera, contains tetrodotoxin (the same neurotoxin that occurs in the venom glands of the blue-ringed octopus), which is lethal if ingested. First aid measures include the administration of an emetic in the early stages of poisoning, and, if paralysis of the respiratory muscles occurs, prolonged artificial respiration. See **fugu** and **porcupinefish**.

**The banded toadfish *Torguigener pleurogramma* grows to about 22 cm in length and is common in estuaries on Australia's east and west coasts.**

**pulmonary barotrauma.** Tissue damage of the lungs resulting from the expansion or contraction of the enclosed air spaces. This is the clinical manifestation of Boyle's Law. There are several types of pulmonary barotrauma including:

• Pulmonary Barotrauma of Descent (lung squeeze)

In breathhold diving the lung volume decreases on descending (Boyle's Law). Once the air in the lungs is equal to the residual volume, lung compressibility ceases and further descent results in haemorrhage, oedema and pulmonary congestion. First aid includes the administration of oxygen. Lung squeeze should not occur in scuba divers as the inhaled air should be at the same pressure as the surrounding water.

• Pulmonary Barotrauma of Ascent (burst lung)

If a scuba diver holds his/her breath while ascending the lung volume increases, causing rupture of the lungs. This can occur in very shallow water, as little as one metre in depth. It can be avoided by breathing normally during ascent. As a result of a burst lung, one or all of the following could occur in any combination:

• Air Embolism

This is potentially the most serious of diving accidents. Serious effects may result from air entering veins and arteries from a ruptured lung. If bubbles of air pass into and block the vessels in the heart or brain, death can follow rapidly. Treatment is urgent and the patient should be moved to a recompression facility as soon as practical. First aid measures include lying the patient on their left side with the feet elevated and the head down. Administer 100% oxygen at atmospheric pressure.

• Pneumothorax

The bag or sac containing the lungs is called the pleura. If the lung is ruptured, air enters the pleural cavity and causes the lung to partially or fully collapse. First aid and treatment as for air embolism.

• Mediastinal and Subcutaneous Emphysema

After the alveoli sacs in the lung rupture, air may escape into the tissues surrounding the blood vessels and airways, sometimes finding its way deep into the chest cavity (mediastinal emphysema). The air may even work its way up to the neck region and lodge under the surface tissues causing a crackling sound when the skin is touched (subcutaneous emphysema). 100% oxygen should be administered at atmospheric pressure and the patient observed at all times until he/she can be taken to a diving medical facility for assessment of their condition.

**purge button.** A button on the second stage (mouthpiece) of a regulator, which when depressed causes a rush of free flowing air that clears water and sometimes sand from the regulator.

**Depressing the purge button releases a flow of air which clears the regulator.**

**quarantine.** The isolation of aquarium fishes, invertebrates and plants that have been captured from the wild or imported before they are introduced into an established aquarium. This procedure helps avoid the spread of diseases. It is suggested that four weeks is an adequate period of quarantine. An added precautionary measure for newly acquired fish is treatment in a bath of copper sulphate solution (1 gm/10 litres of water) before adding them to your main aquarium.

**Queensland groper.** Large fish and member of the family Serranidae. The Queensland groper *Epinephelus lanceolatus,* is one of the largest reef-dwelling fish in Australian waters growing to three metres in length and about 288 kg (634 lb) in weight. Found along the the entire northern half of Australia's coastline and over a large area of the tropical Indo-Pacific extending from east Africa to the islands of the central Pacific Ocean.

**Queensland gropers gained a bad reputation in the days of the hard-hat pearl divers but this reputation does not seem to be deserved.**

**Queensland National Parks and Wildlife Service.** The Queensland State Government department that is responsible for administering the terrestrial national parks in Queensland as well as the day-to-day running of the Great Barrier Reef Marine Park. The service also helps to monitor the humpback whale population off eastern Australia in conjunction with James Cook University, Townsville. It seeks information about whale migration routes through the Great Barrier Reef. Details of sightings should be sent to: Queensland National Parks and Wildlife Service, PO Box 155, North Quay, Qld 4000; phone: (07) 227 8186.

**radio beacon.** See **EPIRB**.

**radio direction finder.** Special radio and aerial used to take bearings from a navigational radio beacon or a commercial radio station as an aid to navigation.

**radio distress frequencies.** Special radio frequencies called in the event of an emergency at sea. The following radio frequencies are internationally recognised:
- 2182 kHz: this frequency has a range of about 100 nautical miles (greater at night).
- 4125 kHz: this frequency has a range of about 200 nautical miles (greater at night).
- 6215.5 kHz: this frequency has a range of about 500 nautical miles (very long-range at night).
- 156.80 kHz: (VHF channel 16) range is line of sight distance and dependent upon the height of the aerial. This frequency is not normally affected by day/night conditions.

In addition to these frequencies, some yacht and fishing clubs use 27.880 MHz for safety purposes. This frequency has a limited range, but under certain conditions, is subject to 'skip' which results in the signal being received hundreds of nautical miles away but being missed by local stations.

**raptures of the deep.** See **nitrogen narcosis**.

**rate of ascent.** The speed at which a diver returns to the surface from depth. The maximum rate of ascent is 15 metres per minute (RNPL–BSAC tables) or 18 metres per minute (US Navy tables and DCIEM tables), though in practice this is very hard to judge. The latest dive computers have an ascent rate warning. Stop five metres below the surface for three to five minutes before surfacing as an added precaution on all dives. See **dive computer**.

**rate of descent.** The speed at which a diver descends from the surface to depth. The recommended maximum rate of descent is 30 metres per minute (RNPL–BSAC tables) or 18 metres per minute or slower (DCIEM tables).

**rebreather.** See **closed circuit breathing apparatus**.

**recompression chamber.** A cylindrical steel pressure vessel used to treat scuba divers with certain types of adverse diving related conditions by returning them to a high ambient pressure. This specially constructed steel chamber enables the trained operator to increase the pressure until the diver's symptoms are relieved. This increase in pressure causes a decrease in size of any nitrogen bubbles present in the body which, in conjunction with oxygen therapy, helps to remove excess nitrogen through respiration. Recompression should be carried out under strict supervision following a specific set of tables. In an emergency a chamber and medical help can be located by calling one of the telephone numbers listed in this book under Diving Emergency Service. See **Diving Emergency Service (DES)** and **decompression sickness**.

**red alga.** Member of the division Rhodophyta which contains the largest group of large marine algae (seaweeds). These algae contain red pigments that mask the green colour of the chlorophyll and range in colour from red to greenish-brown. Red algae are common below low tide level and the red pigments may help the algae to gather sunlight at depths deficient in red light. In Japan, red algae particularly *Porphyra* spp., are cultivated on hanging nets, collected, dried and used for local consumption as well as export. These algae are commonly called 'nori' by the Japanese. See **nori**.

**The small red alga *Champia* sp. is obvious only due to its brilliant iridescence.**

**red tide.** A phenomenon where, under special climatic conditions in tropical and temperate oceans, certain types of plankton such as species of the dino-flagellates *Gymnodinium* and *Gonyaulax* (and sometimes cyanobacteria) reproduce in such huge quantities that in localised areas the sea turns red (sometimes yellow or brown). Neurotoxins such as saxatoxin produced by these dinoflagellates can kill fishes and are concentrated by filter feeders such as oysters, mussels and clams, rendering them unsafe for human consumption. The human condition which results from eating contaminated molluscs is called paralytic shellfish poisoning or PSP. See **neurotoxin** and **saxitoxin**.

**reef.** A rigid, wave-resistant formation of rock, sand or coral which rises from the ocean floor.

**regulator.** A device which reduces the high pressure air in a scuba cylinder to the ambient pressure, enabling a scuba diver to breathe from a cylinder via the second stage of the regulator at a given depth. The regulator reduces cylinder pressure at the cylinder valve (first stage) and the second stage reduction to ambient pressure occurs at the regulator mouthpiece. The most common regulators used today are single hose two-stage types, with a balanced first stage and a downstream valve on the second stage. See **octopus regulator**.

There is a vast selection of modern scuba regulators available today from a host of manufacturers.

**repetitive dive.** • Any dive performed before a 12 hour surface interval has expired (US Navy dive tables). • Any dive performed before a six hour surface interval has expired if the maximum depth was less than 40 metres or between 6 and 16 hours if the maximum depth was greater than 40 metres (RNPL–BSAC dive tables). • Any dive performed before 9 to 18 hours have elapsed, the exact time depends on the bottom time of the previous dive (DCIEM dive tables).

**repetitive group designation.** A dive exposure guide given by letter according to the bottom times listed on your dive tables; the higher the letter, the greater the diver's exposure to nitrogen. Using the US Navy dive tables if a diver wishes to dive again within 12 hours of the first dive, this letter can be used to calculate the maximum bottom time allowable on the next dive.

**rescue beacon.** See **EPIRB**.

**residual nitrogen.** The amount of nitrogen in the blood and tissues of a diver in excess of the normal basal levels, after completion of a scuba dive.

**residual nitrogen time.** The amount of time which must be added to the bottom time of a repetitive dive to compensate for the residual nitrogen accumulated in the body tissues from the previous dive. The residual nitrogen time can be calculated from the dive tables.

**residual volume.** The air remaining in the lungs after a forceful exhalation.

**rhizome.** A root-like underground stem, usually horizontal, which sends out roots below and shoots from the upper surface. See **Zostera**.

**rip.** A strong current flowing away from the shore, dangerous to the uninitiated and the cause of many drownings. Often seen as a zone of agitated white water among otherwise normal wave patterns.

**RNPL Air Decompression Tables.** Dive tables released in 1975 by the Royal Navy Physiological Laboratory at Alverstoke, England, for use by amateur divers. Thought to be more conservative than the US Navy dive tables. See **air decompression tables**.

**RNPL–BSAC Air Decompression Tables.** A set of air decompression tables produced for the British Sub-Aqua Club by the Royal Naval Physiological Laboratory at Alverstoke, England. See **air decompression tables**.

**rock lobster.** Crustacean and member of the order Decapoda, infraorder Palinura. The rock lobster or spiny lobster is also called a 'cray' or 'crayfish' but the official name is rock lobster. Australia exports large quantities of rock lobsters. The three main commercial species are: the eastern rock lobster *Jasus verreauxi* of southern Queensland, New South Wales and Victoria; the southern rock lobster *Jasus novaehollandiae* of Victoria, South Australia and Tasmania; and the western rock lobster *Panulirus cygnus* of Western Australia. There are restrictions on taking rock lobsters by diving but as the regulations vary from state to state, they are outside the scope of this publication. For a full explanation of the crayfish versus rock lobster terminology please refer to *Australian Seashores* by W.J. Dakin and Isobel Bennett; Angus & Robertson, North Ryde, NSW 1987. See **tropical rock lobster** and **western rock lobster**.

The southern rock lobster *Jasus novaehollandiae* is commercially exploited in Victoria, South Australia and Tasmania.

The eastern rock lobster *Jasus verreauxi* cannot be mistaken for any other species because of its characteristic green colouration.

**roe.** The mass of eggs within the ovarian membrane of a female fish or the sperm or milt of a male fish.

**Ron and Valerie Taylor.** Australian husband and wife scuba diving team renowned for their underwater photographic skills and their passion for the conservation of marine life, especially sharks.

**Royal Zoological Society of New South Wales (RZS).** A Society founded in 1879, interested in the study and conservation of Australian animals. It promotes these interests by various means: the awarding of grants, symposia, wildlife forums, day and weekend excursions and publications. The Society also comprises several smaller sections for those interested in specific fields of zoology such as mammalogy or conchology. Membership in the Royal Zoological Society of New South Wales can open many doors into an understanding of Australia's native fauna as well as provide for participation in the Society's activities, a personal pass for free entry into Taronga and Western Plains Zoo, use of the RZS library, and the Society's quarterly magazine, the *Australian Zoologist.* Membership is open to all persons with an interest in the Australian fauna. Write to the Secretary, Royal Zoological Society of NSW, PO Box 20, Mosman, NSW 2088.

**rumbling.** The rotation of scuba cylinders containing abrasive material to remove internal rust. When scuba cylinders are visually tested each year they are inspected for pitting and loose rust: if present, the cylinders are rumbled, a process involving the introduction of a number of small abrasive objects such as marble chips or steel shot into the cylinder, which is rotated slowly for about 24 hours. This process removes all the loose scale and rust, allowing a proper visual inspection for deep pitting which could result in the cylinder being condemned. Deeply pitted cylinders can explode if pressurised. See **burst disc**.

**Safety Sausage.** An inflatable marine rescue tube made of bright red plastic about 20 cm in diameter which can be stored in the pocket of a BC jacket and in an emergency can be inflated using a regulator mouthpiece, or manually by exhaling into the open end. When fully inflated the Safety Sausage is held by the open end about 40 cm under the water, this action causes the tube to remain upright. In an inflated vertical position it is 2.4 metres in height and visible for at least one kilometre. Horizontal on the water it can be seen from the air at 2000 feet. The Safety Sausage is a handy item to carry with you on all dives especially drift dives where there is a greater chance that the diver may become separated from other members of the group or the dive boat. Talk to your local dive shop or the distributor: Diving Security PO Box 6298, Melbourne, Vic. 3004; manufactured by: R.L. Begg, PO Box 5216, Dunedin, New Zealand.

**The highly visible red plastic inflatable safety tube called the 'Safety Sausage' is very useful when diving in strong currents as support craft can locate missing divers quickly.**

**salp.** Member of the class Thaliacea, order Salpida. A group of floating solitary or colonial planktonic tunicates joined in chains, which look like small blobs of jelly. Siphons are situated at opposite ends of the body and the water current which passes through is used for locomotion, respiration and feeding. Individuals vary in length from a few millimetres to 30 cm. They also have a thick gelatinous tunic and muscle bands which do not form complete rings around the body. Fertilisation occurs internally and eggs develop close to the exhalent cavity of the parent where they are expelled on maturity.

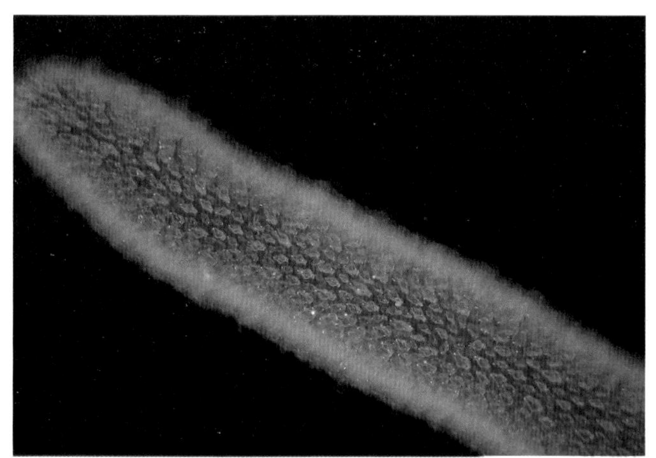

**The salp *Pyrosoma* sp. makes up part of the floating life in the ocean called plankton.**

**Sarcophyton.** A genus of soft corals belonging to the family Alcyoniidae. The colonies are thickened, low in profile and spreading; some species have erect finger shapes while others have wave-like contours on the surface. *Sarcophyton* spp. are very common on Indo-Pacific coral reefs. See **soft coral**.

Soft corals of the genus *Sarcophyton* are very common on Indo-Pacific coral reefs.

**sashimi.** A Japanese dish comprised of thinly sliced pieces of top quality fish, usually salmon, garfish, tuna or even sea-urchin roe, eaten raw with soya sauce, Japanese horseradish and/or ginger. See **roe** and **sushi**.

**saturation diving.** Diving in situations where the diver's body tissues are completely saturated with nitrogen, and decompression time is at a maximum; at this point saturation does not increase with prolonged exposure. Professional divers are sometimes required to work many hours at depth and in order to work efficiently they live in an underwater habitat, or on the surface in a pressurised chamber and are transported to the dive site for regular periods of work. When the job is completed they undergo one long slow decompression in the safety of a pressurised container at the surface.

**saxitoxin.** One of the most potent marine neurotoxins known and the cause of paralytic shellfish poisoning. When a red tide occurs, filter feeders such as oysters, clams, mussels and scallops may consume dinoflagellates (probably species of *Gonyaulax*) containing this powerful neurotoxin which is stored in their tissues. If these contaminated shellfish are eaten the toxin can cause death by paralysis of the respiratory muscles. See **neurotoxin** and **red tide**.

**scallop.** Important commercial bivalve mollusc and member of the family Pectinidae. In Queensland and Western Australia the saucer scallop, *Amusium balloti,* (also called Sun and Moon scallop) is fished commercially. The common species fished from New South Wales estuaries is *Pecten fumatus,* and in Tasmania the commercial species is known as *Pecten meridionalis*. Scallops have an interesting swimming action and move about by opening and closing their shells rapidly and jetting water in short rapid bursts to escape predators such as sea-stars.

**The king scallop *Pecten meridonalis* forms the basis of the scallop industry in Tasmanian waters.**

**The Sun and Moon scallop *Amusium ballati* is fished commercially in Queensland and Western Australian waters.**

**scientific classification of marine organisms.** A hierarchical system of naming and grouping organisms necessitated by the great diversity of plants and animals in the world. Animals or plants constituting a species can loosely be described as a group of individuals capable of interbreeding, but which cannot breed with individuals from another such group. Each species is given a Latin name consisting of two parts, the first (generic) part of the name is shared by closely related species. The second (specific) part of the name refers only to that particular species. This binomial system of naming is in universal use for all living things. Just as several species can be grouped into a genus, several genera can be grouped into a family and several families into an order and so on. The principal divisions within the phylogenetic classification of the Animal and Plant Kingdom are: Phylum (in the Animal Kingdom) or Division (in the Plant Kingdom), Class, Order, Family, Genus, and Species. The first (generic) part of the name is written with an initial capital letter while the second (specific) part is written with a small initial letter. The genus and species should always be printed in *italics* or underlined if *italics* are not available.

**scientific diving.** Diving for the purpose of gaining scientific knowledge by weighing, measuring, examining, photographing and/or collecting physical oceanographic data, minerals or aquatic plants and animals. Scientific divers are usually biologists, chemists, archaeologists, geologists or oceanographers who are employed mainly by federal or state research organisations to collect data for various research projects. See **Australian Scientific Divers Association**.

**Scientists exploring the cold, tannin stained waters of Port Davey in south-west Tasmania.**

**Scleractinia.** The order of the coelenterates containing the hard or stony corals. See **hard coral**.

**scombroid poisoning.** Poisoning that can occur when mackerel-like fishes such as tuna, albacore or bonito are eaten after being left in the sun, or at room temperature for too long. Histidine in the muscle tissue is changed by bacterial action into a histamine-like substance called saurine, which if ingested causes a series of symptoms including; hypotension, cardiovascular shock, rapid weak pulse, bronchospasm and common cold type symptoms.

**scrimshaw.** Handcarved or scratched artifacts made of whale bone, ivory, steel or wood. In the days of the tall ships these artifacts were made by sailors in their leisure time on the long journey back to port. Scrimshaw included: combs, forks, ladles and model ships. A common article carved from bone was the 'jagging wheel', a gadget consisting of a wheel and handle used for decorating pastries. All manner of scrimshaw including intricately carved whale teeth and walrus tusks were given to loved ones or sold on return to port. *Scrimshaw* is also thc official newsletter of Project Jonah. See **Project Jonah** and **whale products**.

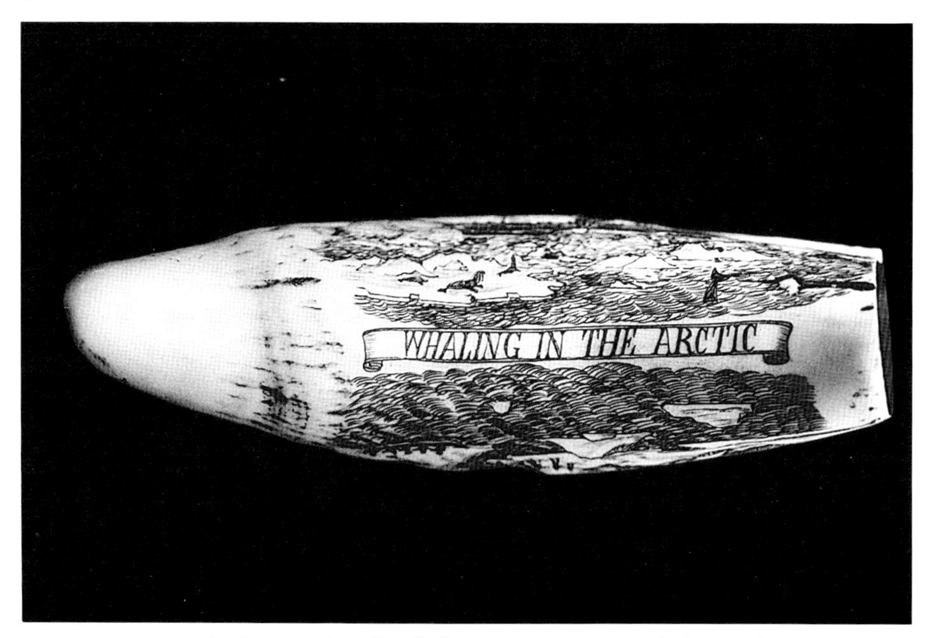


**A scrimshaw carving of a whaling scene on a sperm whale's tooth.**

**Scripps Institution of Oceanography.** World famous oceanographic institute situated at La Jolla, California, USA and part of the University of California. Contact address: University of California-San Diego A-033B, La Jolla, CA 92093; phone: (619) 534 1294.

**scuba.** An acronym of self-contained underwater breathing apparatus.

**scuba cylinder.** A cylinder used to store air under great pressure and from which scuba divers breathe underwater with the aid of a regulator which controls the pressure. Scuba tanks are constructed from either steel or aluminium in many different capacities. The size or capacity of a scuba cylinder refers to the equivalent volume of air at atmospheric pressure held in the cylinder when filled to the maximum recommended pressure. It is debatable whether a steel or aluminium cylinder is best. It should be noted that aluminium cylinders are more buoyant at the end of a dive than the equivalent capacity steel cylinder. The new generation steel scuba cylinders use special light-weight yet strong chrome-molybdenum steel which can be used at pressures up to 302 bars (4500 psi). This allows the manufacture of smaller and lighter cylinders having air capacities similar to larger conventional steel or aluminium cylinders. An Australian company, CIG, manufactures aluminium cylinders for domestic use and for export. Cost may be the major consideration for the choice of a scuba cylinder in the future.

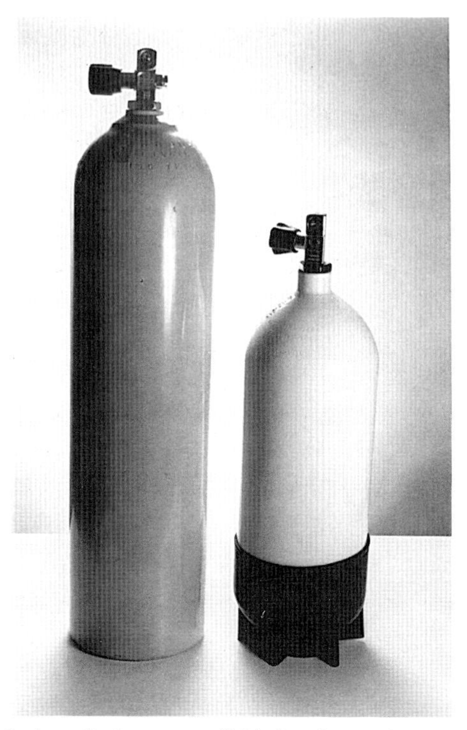

**Scuba cylinders are available in a large selection of capacities. Aluminium cylinders are manufactured with flat bottoms whereas steel cylinders have round bottoms and require a tank boot to hold them upright.**

**scuba cylinder code and markings.** Scuba cylinders should have letters and numbers stamped on the upper portion of the neck which indicate: date of manufacture, what Standard was followed in the manufacturing process (e.g. AS1777 refers to aluminium tanks manufactured to Australian Standards code 1777), serial number, working pressure, test pressure, water capacity, tare weight and test history. For an overview of these markings please refer to the book *The Magic of Scuba* by Bill Sylvester; Wedneil Publications, Victoria 1987.

**scuba cylinder testing.** A testing procedure to check on the condition of scuba cylinders. Full hydrostatic testing of aluminium and steel scuba cylinders is required annually in Australia (Australian Standard AS 2337.1-1989).

**scuba deep diving record.** John J. Gruener and R. Neal Watson of the USA dived to 133 metres (436 feet) off Freeport, Grand Bahama on the 14 October 1968. The women's record of 105 metres (344 feet) is held by Marty Dunwoody of the USA; she completed the dive on the 20 December 1987 off Bimini, Bahama Islands.

***Scuba Diver* magazine.** See **dive magazines**.

**scuba equipment.** Diving equipment used by scuba divers. The basic equipment needed for safe scuba diving is listed:

- Mask
- Snorkel
- Fins
- Knife
- Weightbelt with weights
- Buoyancy compensator
- Scuba cylinder with regulator
- Contents gauge or combo gauge
- Depth gauge
- Watch
- Wetsuit or drysuit

Gloves, compass, torch and dive slate are optional. It is also advisable to have an extra second stage (regulator mouthpiece) fitted to the first stage of your regulator to enable buddy breathing to be performed more easily. This is commonly called an octopus regulator and many dive charter operators insist on divers having one. See **octopus regulator**.

**Scuba Schools International (SSI).** A scuba certification agency formed in California in 1970. The following information was supplied by SSI Australia:

Scuba Schools International is an international scuba certification agency with a business philosophy of customer service that is focused on providing retail dive stores with state of the art educational business and support programs.

SSI started in California in 1970, filling a need that wasn't met by the other scuba certification agencies. Over the years SSI has grown to one of the top three scuba certification agencies in the world. It is also a professional dive store retail association that is dedicated to providing through these member stores, top quality scuba certification training and materials. It is only through these member stores that consumers can become SSI certified, and purchase products. This is the most significant difference between SSI and other scuba certification agencies. SSI heavily supports the diving retailer.

SSI offers a full range of diving courses, from snorkelling and introductory courses, to updates and advanced training. SSI is the first agency to incorporate a combination of training and logged dives into the requirements of advanced training. SSI offers advanced training in several areas including boat diving, night/limited visibility diving, dry suit diving, stress and rescue and many more. These specialty courses, along with actual logged dives, can be combined to meet requirements for all levels of certification.

SSIs Total DiveLog System, a comprehensive and long-term logbook, is a perfect record keeper for such specialty training. It has recently been recognised by *Skin Diver* magazine as the 'best in diving today' and given the Gold Medal Award for excellence.

SSI is also the first and only scuba certification agency to incorporate full-motion video into the teaching system. Video is used in all SSI training programs. The Open Water Diver Training System, which incorporates a home study version enabling students to structure a personal certification schedule, popularised this concept. It is another product from SSI that has received Skin Diver's Gold Medal Award.

SSI Australia opened its doors on 1 January 1989. Since that time it has grown to encompass some 350 professional members and 55 member stores. The philosophy of unsurpassed service and the highest quality products and training programmes has ensured this growth. A reflection of this dedication to service was the introduction of Ski and Sea travel to meet the needs of the travelling diver and member stores ability to offer a diverse product range.

In 1992 SSI Australia released MedDive. MedDive is aimed at the professional instructor, assistant and divemaster and provides training in the on-site handling and treatment of diver related injuries. For more information contact: SSI International Australia, PO Box 662, Balgowlah, NSW 2093; phone: (02) 907 0322, fax: (02) 907 0416 or SSI International, 2619 Canton Ct., Ft. Collins, CO 80521 USA; phone: (303) 482 0883, fax: (303) 482 6157.

**sea.** The salt water covering four fifths of the earth's surface. Also the immediate effect of the wind on the water surface. The word 'sea' is the name given to waves generated by local land and sea-breezes. When sea waves leave the generating area (the area in which waves are formed) they become swell. The height of sea waves is dependant on the strength and duration of the wind in the generating area, and the fetch or distance over which the wind has been blowing. Also used as a term to describe conditions i.e. a big sea—large waves/heavy conditions; slight sea—small waves/good conditions etc. See **swell**.

**sea-anchor.** A floating anchor consisting of a long length of rope, a bucket, or a framed cone of canvas which is trailed from either a ship's bow or stern to give steerage and minimise drifting at sea. A sea-anchor can keep the bow of a disabled ship pointed into the wind to keep it from broaching. This device should be part of the safety equipment carried by all boats operating in open waters. A polyurethane high-tech version of the sea-anchor has been invented and is called a Seabrake. It performs the above functions, but is specifically designed to be trailed

**The Seabrake is a high-tech version of a sea-anchor.**

from the stern when there is danger of broaching or pitchpolling if the boat's speed cannot be controlled by reducing engine revolutions alone. It can also be used for emergency steering, giving better directional stability in following seas or when towing other vessels. The Seabrake is available in five sizes, and custom made versions are also available from Seabrake International, Western Australia or at your local boating dealer. See **broach** and **pitchpolling**.

**sea-anemone.** A coelenterate and member of the order Actiniaria, sometimes referred to as a 'flower of the sea'. Anemones consist of a short column of jelly-like material attached to the substrate by a circular suction disc. On top of this column is another disc with a central hole that functions as both mouth and anus. The mouth is surrounded by a number of tentacles which gather food particles from the passing water and transfer them to the mouth parts for digestion. Anemones can be successfully kept in seawater aquariums. See **swimming anemone**.

**The waratah anemone *Actinia tenebrosa* is viviparous and the young anemones are born through the mouth at the base of the tentacles. Recent scientific studies indicate that these anemones live for at least 50 years.**

**Seabrake.** See **sea-anchor**.

**sea-cucumber.** Echinoderm and member of the class Holothurioidea. The edible species are known as trepang or beche-de-mer. Sea-cucumbers have mouths ringed by tentacles that are used to move sediment into the gut where the organic material is digested and the remaining sand and/or mud passes through unchanged. Sometimes the digestive tract is host to a long slender transparent fish which can at times be seen at the entrance to the anus. Sea-cucumbers range in size from three centimetres to over one metre in length, depending on the species. When removed from the water they should be handled carefully because the sticky white viscera they expel can cause blindness if rubbed into the eyes. The Chinese have eaten trepang for centuries and in Asia it is usually purchased in a dried or smoked form which is then soaked in water and cooked in many different ways. Trepang are rich in protein (43%) and low in fat (2%) and formed the basis of one of Australia's early export industries: exports of trepang to Hong Kong and China began in 1874 and continued until the end of the Second World War.

**The red prickly holothurian *Thelenata ananas* is common in Great Barrier Reef waters.**

**sea-dragon.** A highly modified relative of the sea-horse, unique to Australia and member of the family Syngnathidae. Sea-dragons differ from sea-horses in having leafy appendages and lacking a tail that can be coiled up. The two common species are the leafy sea-dragon, *Phycodurus eques,* found off the coast of South Australia and southern Western Australia and the common sea-dragon, *Phyllopteryx taeniolatus,* found in waters off New South Wales, South

Australia and Tasmania. For more information see fish facts in the April/May 1991 edition of *Scuba Diver* magazine. See **sea-horse**.

The common sea-dragon *Phyllopteryx taeniolatus* grows to 45 cm in length.

**sea-fan.** Horny coral and member of the subclass Alcyonaria or Octocorallia, order Gorgonacea. See **gorgonian**.

**sea-gooseberry.** See **ctenophore**.

**sea-grass.** Marine flowering plant (Angiosperm). There are about 11 genera and 26 species of sea-grasses living in Australian waters. In the relatively sheltered areas of the Australian coastline, such as bays and estuaries, sea-grasses make up a large part of the marine vegetation. They are characteristically green in colour and at certain stages of their life cycle produce inconspicuous flowers and set seeds as do flowering land plants. *Zostera muelleri* or eel-grass is found throughout the world and is common along the south-eastern Australian coastline. The strap-weed, *Posidonia australis,* also occurs in this area. Paddle weed, *Halophila ovalis,* is more common in warmer waters, while *Amphibolus antarctica* occurs only in cold water, from Victoria to Western Australia. When sea-grasses die and rot away they add to the organic matter present in the sediment and become food for worms and crustaceans which are in turn eaten by fishes. The strands of fibre left behind after sea-grasses decompose are sometimes rolled along the ocean floor to form fibre balls and become beachcombers' curiosities when they are washed up on shore after storms. Sea-grass beds are the breeding grounds of many of our commercial species of fishes and invertebrates such as various species of prawns and crabs. These environmentally sensitive sea-grass meadows have been devastated in certain areas by land fill operations for commercial developments such as housing estates, resorts and marina complexes, and by pollution of our marine environment by industrial waste products and oil spills. See **eel-grass**.

**The strap-weed *Posidonia australis* provides shelter and protection for a myriad of juvenile fishes and crustaceans.**

**The stringy fibrous material remaining from the decomposition of sea-grasses can sometimes form fibre balls.**

**sea-hare.** A gastropod mollusc and member of the subclass Opisthobranchia, order Anaspidea. Sea-hares are the largest of the opistobranch molluscs, and have an internal shell (or none at all). They are herbivorous, feeding mainly on algae and are commonly found in large numbers at certain times of the year on the sea-grass beds of shallow estuaries in most Australian States where they congregate to lay their long strings of eggs. The mantle cavity contains a gill for respiration and a gland containing a purple dye which is expelled into the water when it is handled roughly or attacked by a predator. It is easy to understand why these animals are referred to as sea-hares, as they are rabbit or hare-like with a domed shaped back and erect ear-like appendages on the head.

**This sea-hare genus *Aplysia* is grazing on fine filamentous algae.**

**sea-horse.** A species of fish and member of the order Syngnathiformes (pipefish and sea-horses), family Syngnathidae. Sea-horses swim in an upright position by means of a wave-like motion of the dorsal and pectoral fins. The female of the species lays her eggs in the male's brood pouch. The male incubates the eggs until they hatch about 20 days later. Sea-horses are common aquarium fishes and survive well on a diet of brine shrimp. White's sea-horse *Hippocampus whitei* is a common species found in most Australian States. For more information on sea-horses see Fishwatcher's Notebook in the April/May 1987 issue of *Scuba Diver* magazine. See **sea-dragon**.

**White's sea-horse *Hippocampus whitei* is variable in colour from bright yellow to brown and grows to about 20 cm in length.**

**sealer.** A person or ship engaged in hunting seals.

**sea-lettuce.** A green alga and member of the division Chlorophyta. See **Ulva lactuca**.

**sea-lion.** Marine mammal and member of the family Otariidae. See **Australian sea-lion**.

**sea-moss.** See **bryozoan**.

**sea-mouse.** Annelid worm and member of the class Polychaeta, family Aphroditidae, a group of marine annelid worms which are bulky and flattened in appearance and grow to 15–20 cm in length. Their fine bristles form a felt-like covering which is interspersed with thicker stiff bristles that can cause irritation to human skin when touched. They burrow through soft sediments and feed on various invertebrates including other polychaetes.

**sea-pen.** Soft coral and member of the subclass Alcyonaria or Octocorallia, order Pennatulacea. Sea-pens live on soft mud or sandy substrates mostly at depths below 20 metres; occasionally specimens can be found in as little as five metres of water. They are normally erect and pencil or club-shaped, closely resembling a quill pen from which their common name is derived. New polyps are formed by budding from the sides of the older polyps. Sea-pens will live in aquaria if the sand is very fine and at least 9 to 12 cm deep. See **soft coral**.

**The sea-pen *Sarcoptilus* sp. gives off a brief flash of luminous green light when stroked lightly at night.**

**sea-purse.** The horny egg case of various sharks and rays such as the Port

Jackson shark and the Melbourne skate. The whale shark lays an egg case up to 30 cm in length.

**The egg case of the Port Jackson shark.**

**The egg case of the Australian swellshark *Cephaloscyllium laticeps,* also called the draughtboard shark, which lives in southern Australian waters.**

**search and rescue diving.** Diving operations involving searching for persons missing at sea and the recovery of bodies and/or primary and secondary forensic evidence.

**seasickness.** Nausea caused by the motion of a vessel at sea.

**sea-slug.** See **nudibranch**.

**sea-snake.** A marine air-breathing reptile and a member of the family Hydrophiidae. At least 32 species of sea-snakes are found in Australian waters. They are similar in appearance to terrestrial snakes, except for a flattened paddle-like tail. Like some terrestrial snakes they are highly venomous. The beaked sea-snake, *Enhydrina schistosa,* is considered one of the most dangerous to humans, with venom about twice as toxic as Indian cobra venom and about eighty times as toxic as sodium cyanide. Sea-snakes are rear-fanged snakes and the injection of venom is from fangs positioned at the back of the mouth. This, along with the generally small sized head of most sea-snakes encountered, makes it unlikely that they could inject venom into a large creature such as a diver. The fangs are short and face backwards, further reducing the chance of penetrating a diver's five millimetre neoprene wetsuit. Sea-snake bites are almost painless, and no symptoms of envenomation develop in the first thirty minutes to about three hours. The main effects of the venom appear to be associated with damage to the voluntary muscles. Death may occur several hours to a few days later. The Australian Commonwealth Serum Laboratories has developed an antivenom which works with all Australian sea-snake bites, so if you are bitten, exact identification of the offending snake is not necessary. Important adaptations these animals have for living in the sea include: a paddle-like tail to aid in swimming, and a special salt gland just below the sheath encasing the tongue which eliminates excess salt from the blood. Sea-snakes eat predominantly bottom dwelling fishes and eels which are hunted mainly at night; some species however feed exclusively on fish eggs. Sea snakes also have the ability to remain submerged for up to two hours and can dive to about 200 metres to obtain their food if necessary. On the Great Barrier Reef the most common species seen by divers is the olive sea-snake, *Aipysurus laevis,* while in colder waters further south the yellow-bellied sea-snake, *Pelamis platurus,* is the most common species. Predators of sea-snakes include sharks and predatory sea birds.

**The olive sea-snake *Aipysurus laevis.***

**sea-spider.** A pycnogonid and member of the phylum Chelicerata, class Pycnogonida. The sea-spiders or pycnogonids are exclusively marine spider-like animals, similar in appearance to terrestrial spiders but are not true spiders (their relationship to arachnids is unclear). Sea-spiders are usually only a few millimetres long, however some of the deepwater species (6000 metres) have a leg span of around 50 cm. The body of pycnogonids is reduced about as far as it is possible—they are almost all legs and some of the organs which are normally found in the body reside in the first joints of the legs. The eyes, usually four in number, are grouped together on a short stalk. The female sea-spider lays the eggs. The male fertilises them and carries them around on two especially modified legs until they hatch. The larvae moult several times before reaching adult size. Sea-spiders feed mainly on the tissues of coelenterates and bryozoans which they suck up, with the aid of a tubular proboscis. See **proboscis**.

**sea-squirt.** Member of the subphylum Urochordata, class Ascidiacea. See **ascidian** and **tunicate**.

**sea-star.** Echinoderm and member of the class Asteroidea. Radially symmetrical echinoderms, usually with a star-shaped body having five or more arms extending laterally from the central disc. On the lower side of the arms is a furrow from which protrudes a large number of tube feet, the primary means of locomotion.

**The sea-star *Asterodiscides truncatus* has five stubby arms and prominent rounded lumps or tubercles covering the entire dorsal surface.**

Sand dwelling sea-stars burrow into the substrate with the aid of pointed tube feet while species inhabiting rocky bottoms possess suckered feet for climbing and for holding prey. The mouth is situated on the underside or ventral surface while the anus is on the upper or dorsal surface. Sand or mud-dwelling sea-stars are usually omnivorous, feeding on worms, molluscs, micro-organisms and detritus and the large predatory sea-stars such as *Coscinasterias calamaria* feed on large molluscs such as oysters, mussels, scallops and abalone. Most other species feed on encrusting animals such as ascidians, bryozoans and sponges which live mainly on rocks or coral. Sea-stars have the remarkable ability to regenerate lost or damaged arms and some species actually tear themselves in half and regenerate the missing portions using this as a method of reproduction. See **crown-of-thorns starfish**.

**sea-urchin.** Echinoderm and member of the class Echinoidea which also includes sand-dollars, and heart-urchins. Sea-urchins, sometimes referred to as sea-eggs, are generally ovoid in shape and their body walls are composed of hard limey plates that fit together like a jigsaw puzzle. These plates carry long and/or short spines which give the animals a porcupine-like appearance. Some urchins such as the sand-dollars, are very flat and have a covering of very tiny tightly packed spines. Sea-urchins are mainly bottom dwellers and live on both hard and soft substratum, feeding mainly on algae and detritus. Several species of sea-urchins have poisonous spines and/or pedicellariae which can cause painful stings. The pedicellariae of the tropical sea-urchin, *Toxopneustes pileolus* (sometimes called the flower urchin) have three needle-sharp fangs at their tips which can inject venom and cause severe pain. In Japan several deaths have been recorded from drowning after divers have been stung by this species. See **echinoderm** and **pedicellariae**.

**The conspicuous white and mauve sea-urchin *Pseudoboletia indiana* is often found covered with shells and pieces of algae, possibly to help camouflage it from large predators such as wrasses.**

**sea-vegetable.** Edible marine algae commonly called seaweed. Some of the common edible marine algae found in Australian waters belong to the following genera: *Caulerpa, Enteromorpha, Porphyra,* and *Ulva.* Sea-vegetables are used mainly to enhance the flavour of soups, salads and rice dishes. See **nori**.

**sea-wasp.** A cuboid jellyfish with a fatal sting and member of the order Cubomedusae. See **box jellyfish**.

**seaweed.** See **algae**.

**sea-whip.** A horny coral and member of the subclass Alcyonaria or Octocorallia, order Gorgonacea. Sea-whips have a central strengthening rod composed of a very strong and flexible horny material called gorgonin. The rod is attached to a hard surface by a plate or a tuft of creeping branches. See **gorgonian**.

**Sea-whips are firmly attached to the bottom by a tuft of creeping branches.**

**second stage.** See **regulator**.

**seine.** A type of fishing net that hangs vertically in the water; the bottom edge is weighted and the upper edge supported by floats.

**servo-assisted valve.** A demand valve that incorporates both downstream and tilt valves in the second stage of the regulator. This system can be adjusted to provide easy inhalation but is prone to free flowing when the regulator is not in the diver's mouth.

**shagreen.** The dried untanned skin of sharks, originally used as a sandpaper substitute for finishing timber and polishing marble. The Japanese also used shagreen on sword handles, as it provided a good grip, even with hands covered in blood.

**shark.** A fish and member of the class Chondrichthys (cartilaginous fishes), subclass Elasmobranchii. Living sharks are divided into eight orders, each recognisable by certain physical characteristics. Each order is divided into families. There are thirty families and about 344 species of sharks known to science; of these about 90 species occur in Australian waters. The three types regarded as particularly dangerous to humans include the white pointer, tiger shark, and various whaler sharks. Sharks have good eyesight, even in low light conditions and can detect one part blood in 10 million parts water by smelling the water as it passes through their nostrils, which are located under the snout. Their inner ears are sensitive to vibrations travelling through the water and it is for this reason sharks are attracted to the death throes of a fish on the end of a spear or fishing line. In Australia two species of sharks are protected by law. They are the Herbst nurse shark and the grey nurse shark. See **ampullae of Lorenzini** and **specific shark species** e.g. **tiger**, **whaler**, **zebra**, etc.

**A set of jaws from two dangerous species of sharks common to Australian waters.**

**shark attack.** See **Australian Shark Attack File**.

**shark billy.** A short stick with or without a nail in the end, used for fending off sharks. The stick is not used to attack the shark but merely to push it away if it comes too close to the diver.

**shore diving.** Land based diving operations. See **entering** and **exiting the water**.

**signals.** See **underwater communication**.

**silver gull.** A seabird commonly known as a seagull. The silver gull *Larus novaehollandiae* is an important scavenger along the Australian coastline and has spread to inland water courses. They are also found in New Zealand and South Africa.

**The silver gull or 'seagull' *Larus novaehollandiae*.**

**single dive.** The definition of a single dive varies with different dive tables. Using the US Navy tables, a single dive is a dive conducted after a 12 hour surface interval. Using the RNPL–BSAC tables, a single dive is a dive commenced after a six hour surface interval if the maximum depth was less than 40 metres, or any dive after 16 hours if the maximum depth was greater than 40 metres. For the purpose of using air decompression tables all divers should be aware of these definitions.

**single hose regulator.** A regulator that has a single small diameter hose leading from the first stage of the regulator (part attached to the cylinder) to the second stage valve or mouthpiece. Most regulators sold in Australia over the last 25 years are of the single hose variety as opposed to the twin hose variety. See **twin hose regulator**.

**sinus.** One of a number of air-filled hollow cavities mainly in the facial bones, which connect with the nasal passages. If a diver has any congestion in the sinus cavities it is better not to dive as it may not be possible to equalise pressure, resulting in ruptured blood vessels in the sinus cavities which could cause further congestion and possible infection.

**skin barotrauma of descent.** See **suit squeeze**.

**skindiver.** An American term meaning snorkeller; that is a person diving with only the aid of fins, mask and snorkel. In general this term has been used to describe all forms of sport diving.

***skin diver* magazine.** A famous American magazine dedicated to scuba diving, first published in 1952 and still published monthly by Petersen Publishing Company, 8490 Sunset Blvd, Los Angeles, California 90069. It is available from selected newsagents throughout Australia. See **dive magazines**.

**slack water.** The period of time when there is little movement in the water between the inflowing and the outflowing tides. Slack water varies between different geographic locations. If a planned dive is tide-dependant check tide tables and if possible seek local knowledge as to the best time to enter the water.

**slate.** See **dive slate**.

**slurp-gun.** A device similar in appearance to a syringe (though much larger), consisting of a large hollow clear plastic open ended cylinder and an attached plunger. A slurp-gun is used to collect whole, undamaged fishes and other small marine creatures which are sucked into the tube by pulling back rapidly on the plunger. On some models the plunger is activated by large rubber bands and a trigger mechanism.

**snorkel.** Breathing tube for breathhold diving, consisting of a hollow plastic or rubber tube with a U-shaped curve and an attached mouthpiece on the curved end. The snorkel is one of the least complicated and most important pieces of skin diving equipment. Using a snorkel makes it possible to breathe continually while lying face down in the water.

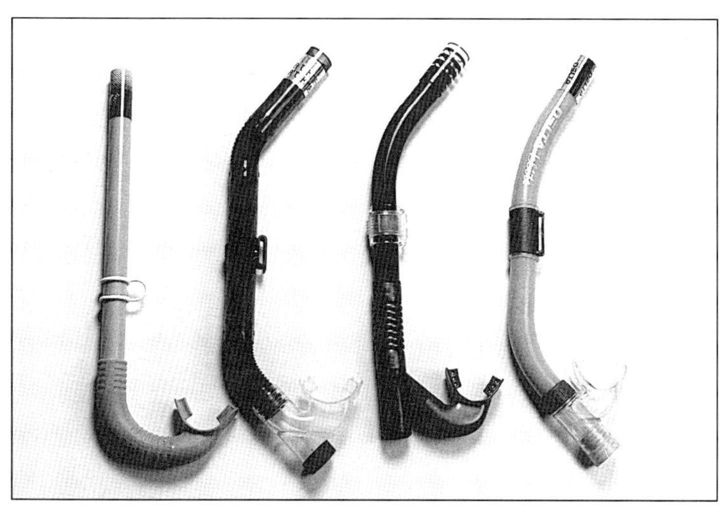

**A selection of snorkels.**

**Society for Underwater Historical Research.** A South Australian based society that surveys and researches shipwrecks. For more information contact: PO Box 181, North Adelaide, SA 5006.

**soft coral.** Member of the subclass Alcyonaria or Octocorallia. These bottom living colonial animals occur in all oceans, from intertidal levels to the deepest abyss, though they are most abundant in shallow tropical seas where they grow as tree-like branched forms or encrusting, lobed or membranous types. Sometimes called octocorals, these marine invertebrates have bodies made up of many polyps connected by canals and inter-connected by fleshy tissue strengthened by calcareous spicules. Each of the polyps has eight feather-like tentacles which are used for capturing tiny particles of planktonic food and transferring it to the mouth at the centre of the polyp. Many shallow water soft corals contain symbiotic zooxanthellae (dinoflagellates), planktonic algae which convert the sun's

**The soft coral *Isus* sp. has yellow polyps. This specimen was photographed on Sweetlip Reef in the Swain Reef complex, Queensland.**

energy into food such as glucose, glycerol and simple amino acids which are utilised by the host coral. In return the host provides carbon dioxide and other substances which the zooxanthellae require. Soft corals are eaten by nudibranchs, predatory molluscs such as the egg cowrie *Ovula ovum* and by the crown-of-thorns starfish. In some places soft corals are the predominant invertebrate organisms on a coral reef, particularly a reef that has been devastated by the crown-of-thorns starfish. See **gorgonian**, ***Sarcophyton***, **sea-fan**, **sea-pen** and **zooxanthellae**.

**sonar.** See **personal dive sonar**.

**sound.** Sound waves travel faster and further in seawater than in freshwater: 1550 metres/second and 1410 metres/second respectively.

**South Pacific Underwater Medicine Society (SPUMS).** The following information has been supplied by SPUMS:

A Society formed on 3 May 1971, when a group of doctors with a common interest in diving and hyperbaric medicine met in the boardroom at HMAS Penguin, Sydney NSW. Dr Carl Edmonds was elected President and Dr Ian Unsworth, Secretary, of the new Society, the South Pacific Underwater Medicine Society.

The group decided that the aims of the Society should include the interchange and dissemination of information on research, recent advances and clinical data in the broad field of Undersea and Hyperbaric Medicine and Physiology and that the Society should assist in the distribution of information to other interested parties such as the Standards Association, government departments and diving organisations. Other doctors were invited to join this select group of about 20 and membership rapidly grew to about 120. From that beginning it rapidly became apparent that many non-medical people associated with the diving industry had interests in common with the aims of the Society. For these people Associate Membership became available, enabling them to receive the benefits of the Journal and to attend meetings. Now Associates make up about 40% of the membership which stands at around 900. Of these, 150 live in New Zealand, 120 in other overseas countries and the remainder reside, at least temporarily, in Australia.

Education is the prime aim of the Society so that each year a scientific meeting is conducted to which all members are invited to attend. A keynote speaker presents his/her work, which acts as the basis for wide ranging discussions. Other regional meetings may also be held depending on the circumstances. Scientific meetings have been held throughout the Indo-Pacific region, usually in areas where diving is a prominent tourist attraction so that the local medical and diving community can benefit from the presence of such a fund of information. Such venues have included places like the Solomon Islands, Philippines, Vanuatu, Fiji, Maldives, Thailand, Singapore and Papua New Guinea. Much of the information presented at the scientific meetings is published in the quarterly *SPUMS Journal,* the only scientific journal in the world devoted to diving medicine. This Journal started as the

*President's Newsletter* and it contained items on recent advances, accounts of overseas trips, news of members and notification of forthcoming meetings. Under the editorial management of Dr Douglas Walker it has grown and developed into a respected scientific journal with contributions from world authorities in the field of underwater and hyperbaric medicine.

SPUMS remains the only Society which awards an academic diploma to members who have achieved specialist status in the field of diving and hyperbaric medicine. In order to gain this award a member must satisfy the committee that he/she has completed six months full-time training in hyperbaric medicine, has completed the approved training courses and their examinations and has presented a thesis, treatise or paper on an approved subject. To date less than 50 doctors have met these criteria. As more training posts become available then more candidates may become eligible for this award.

SPUMS has been able to make significant contributions to the formation of the Diver Emergency Service (DES) network, the West Australian Task Force in Underwater Diving, the Standards Association and the formulation of the PADI Sports Diving Tables. The Society aims to promote diving safety by the dissemination of information about medical problems and the methods by which such problems may be eliminated, avoided or treated. Hon Sec. SPUMS, Suite 6, Killowen House, St. Anne's Hospital, Ellesmere Rd, Mt Lawley, WA 6050.

**space blanket.** An aluminiumised soft plastic sheet used in first aid procedures. A space blanket is handy to include in your diving first aid kit for use in cases of hypothermia (loss of body heat). When wrapped in a space blanket the patient's own body heat is reflected back, this helps to raise the core temperature. See **hypothermia**.

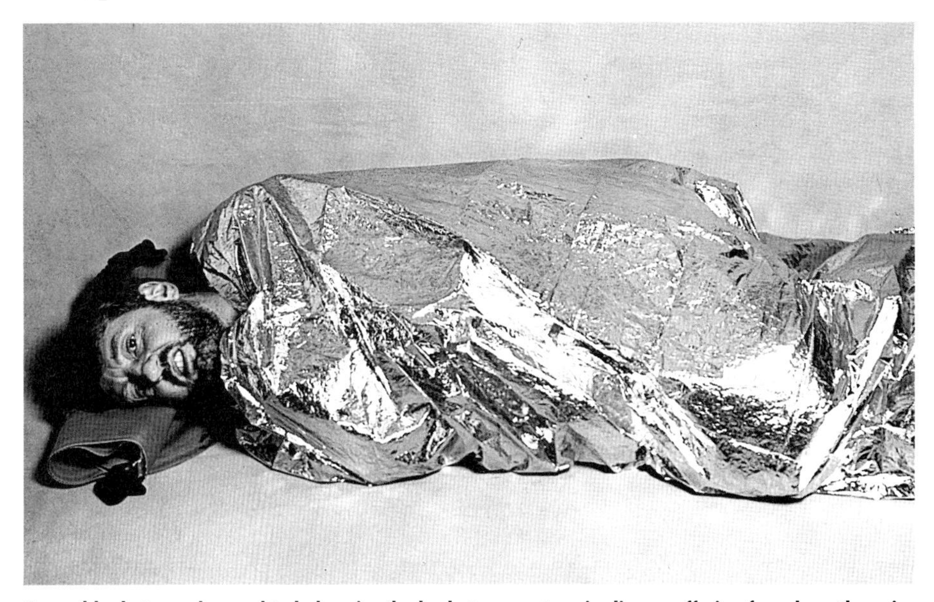

**Space blankets can be used to help raise the body temperature in divers suffering from hypothermia.**

**specialist diving.** See **boat diving**, **cave diving**, **decompression diving**, **night diving**, **scientific diving** and **wreck diving**.

**spermaceti.** A waxy substance that fills a large part of the head of the sperm whale *Physeter macrocephalus.* This substance is composed mainly of cetyl palmitate and is chiefly used as a thickening agent in medical ointments and cosmetic creams. It is also used as an additive in the manufacture of soap, polish and candles. See **whale**.

**spicules.** Calcareous or siliceous supporting rods present in the tissue of sponges, soft corals, gorgonians, holothurians and other marine invertebrates, which help to give shape and rigidity. The presence and shape of spicules is often used by taxonomists as a guide in the identification of species.

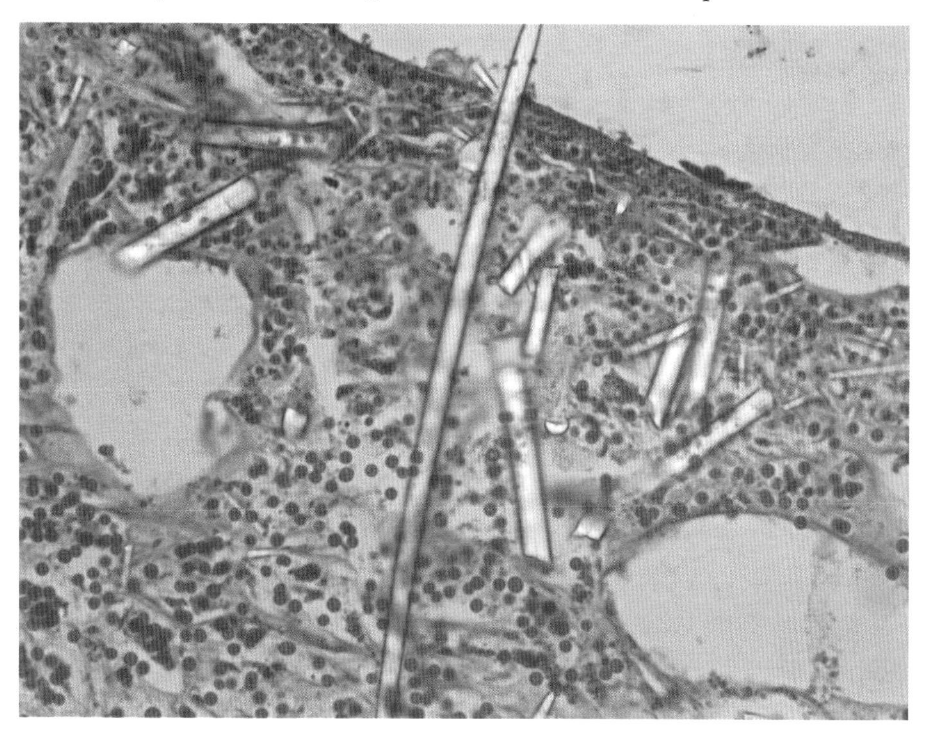

**The clear rod-shaped structures in this photomicrograph of sponge tissue are called spicules.**

**spindrift.** Sea-spray swept by violent winds along the surface of the sea.

**spiracle.** A respiratory opening behind the eye in sharks and rays. See **marine leech for photograph of spiracle**.

**sponge.** The simplest of the multicellular animals and member of the phylum Porifera. The sponge is a sedentary filter-feeding aquatic animal which utilises a single layer of flagellated cells to pump a directional water current through its body. Sponges have extremely variable body shapes and consist of a collection

**The boring sponge *Cliona* sp. is a member of the Demospongiae class of sponges. These sponges are able to excavate complex tunnels in calcareous material such as hard coral and mollusc shells.**

of cells enclosing a system of chambers and canals which are connected to the exterior ocean environment through small openings called pores. These cells are supported by a framework of calcareous or siliceous spicules and/or a fibrous flexible material called spongin. Seawater enters the sponge via the pores, carrying with it a supply of oxygen and food particles, and is passed out via larger openings called oscules. The three major classes of sponges are: class Calcarea (these sponges have spicules composed of calcium carbonate and contain no spongin fibres); class Hexactinellida (these are the deepwater 'glass sponges' with spicules composed of silica and no spongin fibres); and the class Demospongiae (usually with siliceous spicules and/or a fibrous skeleton). The class Demospongiae contains the largest number of sponges and accounts for about 95% of the 5000 or so of the described sponge species. Sponges, being

**The calcareous tropical sponge *Pericharax heteroraphis* contains supporting spicules of calcium carbonate and is a member of the class Calcarea.**

sessile and unable to flee from predators, are thought to have a chemical defence system to enable them to defeat predators and prevent colonisation by other animals. It is for this very reason that sponges are being closely examined by biologists and biochemists as possible sources of pharmaceuticals. See ***Dysidea herbace***, **stinging sponge**, **golf-ball sponge** and **Venus's flower basket**.

***Sportdiving in Australia & the South Pacific.*** A diving magazine first published in 1970, as *Skindiving in Australia.* The magazine had a name change in 1974 to *Skindiving in Australia and New Zealand,* then again in 1980 to *Skindiving in Australia & the South Pacific,* and finally in March 1987 to *Sportdiving in Australia & the South Pacific.* See **dive magazines**.

**spring tide.** The tide occurring at or shortly after the new or the full Moon. When the Sun and Moon are directly in line with Earth, their combined magnetic pull causes the tides to rise to a higher level than normal. During a spring tide high water is higher than normal and low water is lower than normal. Also called a king tide. See **neap tide**.

**SPUMS.** See **South Pacific Underwater Medicine Society**.

**SSI.** See **Scuba Schools International**.

**standard diver.** A hard-hat surface-supplied diver, wearing 'standard dress'.

**standard diving dress.** Equipment consisting of a rigid helmet attached and sealed to a flexible waterproof suit. Air is pumped from the surface to the helmet for the diver to breathe and to pressurise the suit, excess air is bled off through an outlet valve on the helmet. In 1837 Augustus Siebe manufactured his 'closed dress' design diving suit, variations of which are still in use today.

**standby diver.** A fully suited and kitted-up diver ready to enter the water to assist a diver in trouble in the event of an emergency.

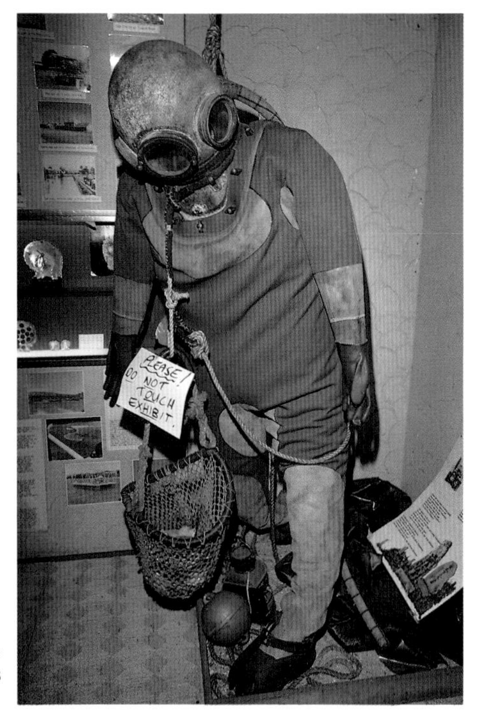


**Standard diving dress was popular in northern Australia for pearling operations early this century.**

**starfish.** See **sea-star** and **crown-of-thorns starfish**.

**stem.** In nautical terms the forward part of a ship.

**stern.** In nautical terms the rear or hinder part of a ship.

**stinger suit.** See **lycra suit**.

**stinging sponge.** Various species of marine sponge, members of the phylum Porifera. The stinging sponge *Neofibularia mordens* is a toxic marine sponge, class Demospongiae, order Poecilosclerida and is endemic to Victorian and South Australian waters. All sponges are toxic to humans, however it appears that the slime from this sponge in particular can cause severe pain and inflammation. Several divers who helped in the collection of this sponge from Port Phillip Bay Victoria for a biological screening programme, were badly effected by the sticky mucus-like slime which exuded from the sponge. The sensation was likened to having thousands of tiny pieces of glass embedded in the skin; touching the skin increased the pain. There was swelling, redness and puffiness of the skin, which continued for weeks, and after a period of three months the skin peeled off one of the diver's feet where the sponge had made direct contact. In tropical waters *Neofibularia irata* is one of several species capable of causing stings and/or rashes when handled. See **sponge**.

**The stinging sponge *Neofibularia mordens* is endemic to Victoria and South Australia.**

**The stinging sponge *Neofibularia irata* of our tropical northern waters should be avoided as it can cause rashes and swelling when handled.**

**Stingose.** A topically applied lotion or gel for marine stings and bites which relieves the pain from the stings of jellyfish, sea-lice and hydroids. Stingose is manufactured in Australia by Hamilton Laboratories and sold by most chemists.

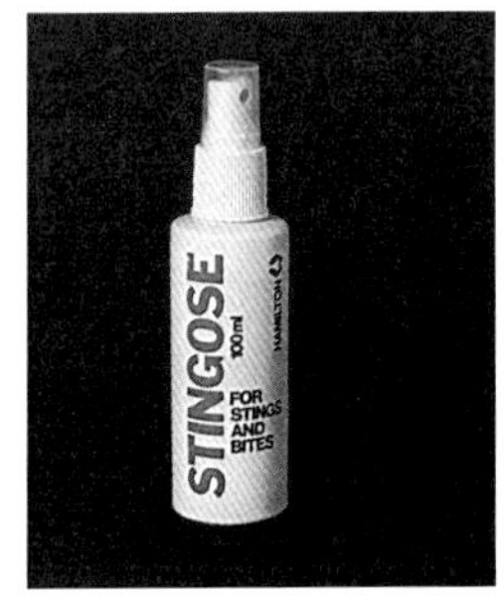


**The topical lotion 'Stingose' can be applied to most marine stings to relieve pain and itching.**

**stonefish.** A deadly fish and member of the family Scorpaenidae. The stonefish *Synanceia horrida,* is a well camouflaged tropical bottom-dwelling fish which has 13 sharp long dorsal spines capable of penetrating the sturdiest boots. The spines are each equipped with two venom sacs, which if trodden on can cause excruciating pain and in severe cases death.

**Tropical divers should be aware of the deadly stonefish *Synanceia horrida* which is sometimes found in considerable numbers inside shipwrecks.**

**storm.** A wind of 48–55 knots or 89–102 km/hr.

**strap-weed.** See **sea-grass**.

**stromatolite.** A marine sedimentary structure built in shallow water by cyanobacteria. The cyanobacteria secrete a sticky mucus which binds fine sediment that over a period of many years accumulates into a dome-shaped stony structure called a stromatolite. Western Australia is the world's mecca for stromatolites and they can be found at the following locations: Hamelin Pool at the southern end of Shark Bay, lakes on Rottnest Island near Perth, Lake Clifton 100 km south of Perth and in Lake Richmond at Rockingham.

**strong breeze.** A wind of 22–27 knots or 39–49 km/hr.

**strong gale.** A wind of 41–47 knots or 75–88 km/hr.

**submersible.** A controlled, free-diving underwater vessel without external ballast. Often used for research purposes.

**subtidal.** Below the level of the tide.

**suit squeeze.** The reduction in volume of pockets of air trapped between the skin and the lining of poorly fitting wetsuits during descent caused by an increase in ambient pressure. The skin tends to be sucked into these pockets, leaving linear weal marks which disappear after several hours on the surface. A common place for suit squeeze to occur is behind the knees.

**surface interval.** The time a diver spends on the surface between dives.

**sushi.** A Japanese delicacy consisting of very thin slices of raw fish or other seafood served with cold, seasoned rice and rolled in a sheet of dried seaweed. See **nori**.

**swell.** Waves that have left the generating area (the area in which waves are formed). Swells are distorted by shallow water, which causes them to steepen and eventually break as surf. A swell may be distinguished by its regularity; sea waves are much more irregular and tend to break at random. Long swells travel faster than short swells, so the first warning of a distant storm may be the presence of a long swell. Often both sea and swell are present, combining to produce high waves.

**swim-bladder.** A gas filled bladder or sac found in many bony fishes in the roof of the abdominal cavity and used to control buoyancy and maintain ease of swimming at any depth. It may also act as a resonating chamber for hearing and sound production.

**swim fins.** See **fins**.

**swimming anemone.** Coelenterate and member of the class Zoantharia, order Actiniaria, characterised by colourful blister-like vesicles covering a bulbous body. Colours vary from crimson or red to brown and green; white or bluish vertical stripes are present on the vesicles. The swimming anemone *Phlyctenactis tuberculosa* grows to about 15 cm in height and is commonly associated with kelp beds, ranging from New South Wales to Western Australia. Unlike most anemones the swimming anemone is unattached at its base and is free to drift along the bottom and attach to anything it bumps into.

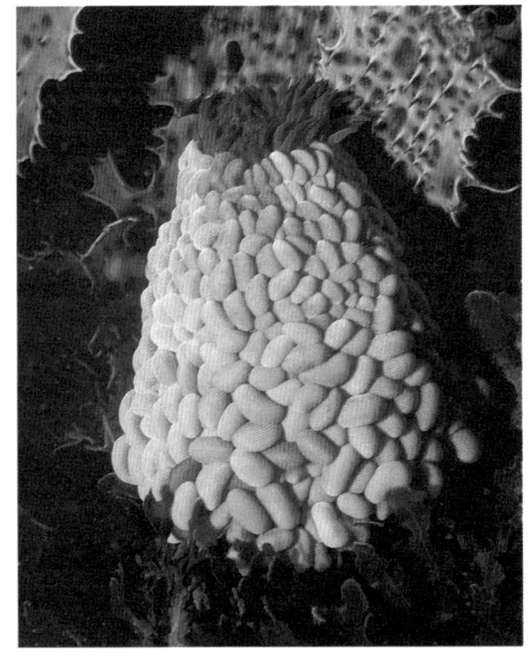

**The swimming anemone *Phlyctenactis tuberculosa* is sometimes marked with two or three longitudinal bands of a contrasting colour.**

**Sydney Aquarium.** An aquarium that is part of Sydney's Darling Harbour Redevelopment, located on the city side of the old Pyrmont bridge. Scuba divers are catered for with 'Diver Awareness Seminars'. These courses teach scuba divers about marine life they may encounter in waters off New South Wales. Special topics will be covered on request. The building housing the Sydney Aquarium was built to emulate a gigantic breaking wave and stands some 14 metres tall and 140 metres in length. The main features are the two floating oceanariums moored alongside the complex. These are the largest floating aquarium tanks in the world, one re-creating the 'open ocean' environment and the other the life below Sydney Harbour. Visitors are transported along a moving footway through an underwater acrylic tunnel which allows a clear view of the many schools of fishes and the larger predators, such as sharks. There are five smaller or micro aquariums. These have video cameras with zoom controls operated by the visitor, allowing magnified views of the smaller occupants, such as sea-anemones, sea-horses and others, on a colour monitor. Other exhibits include; the mangrove swamp habitat, fishes of the far north, the Great Barrier Reef, crocodile and turtle exhibits and the Murray-Darling river system, home of the Murray cod. The Sydney Aquarium is open seven days a week from 9.30am to 9.00pm. Sydney Aquarium, Aquarium Pier, Darling Harbour, Sydney, NSW 2000; phone: (02) 262 2300, fax: (02) 290 3553.

**Sydney Maritime Museum.** The following information has been supplied by the Sydney Maritime Museum:

The Sydney Maritime Museum was begun in 1965, by concerned enthusiasts, with the acquisition of the steam yacht *Lady Hopetoun,* and during its early years was run entirely by member volunteers, who continue to make a tremendous contribution to the Museum's operation even today.

Activities expanded with the restoration of the 1874 three-masted iron barque *James Craig* and today the Museum remains a most important community organisation, independent of governments and strong in the knowledge that it has one of the most important maritime collections in the country and the most significant fleet of operational historic vessels in the Southern Hemisphere. Over the last quarter of a century the Museum has developed a worldwide reputation for authenticity in ship restoration and recording techniques.

As an educational resource the Museum stands out through its active role in reviving ship building and ship repair techniques and through its major collection of books, journals, private papers and maritime artefacts, all significant to New South Wales' maritime heritage.

The Museum presents its floating exhibits at Sydney Seaport in Darling Harbour. Immediately in front of the Harbourside shopping mall, under the shadow of the Pyrmont Bridge, is the *James Craig* and the old Sydney ferry *Kanangra,* both now used for exhibitions and functions. These two ships dominate Sydney Seaport and are surrounded by other smaller vessels like the plush private yacht *Boomerang*, the former Government VIP launch *Lady Hopetoun,* the small passenger ferry *Protex* and the 1902 steam tug *Waratah.*

Today the Museum is undertaking the momentous task of restoring its fleet to top working condition. In addition to its fleet of vessels the Museum is custodian to a collection of artefacts and an excellent library.

The aims of the Sydney Maritime Museum are: to help safeguard Australia's seafaring heritage; to collect and preserve important objects and artefacts; and to foster interest and fellowship among people who share these aims. The quarterly magazine *Australian Sea Heritage* is an excellent means to these ends. It has encouraged a flow of new members whose first contact has been through its pages. Sydney Maritime Museum, PO Box 140, Pyrmont, NSW 2009; phone: (02) 281 0266. See **Australian National Maritime Museum**.

**Sydney rock-oyster.** Bivalve mollusc and member of the family Ostreidae. The New South Wales or Sydney rock-oyster *Saccostrea cucullata* is also known by oyster growers as *Saccostrea commercialis* and is famous worldwide for its good eating qualities. This oyster extends all the way along the eastern coast of Australia from northern Queensland to eastern Victoria. It seems to grow best in New South Wales estuaries where it takes from three to five years to reach commercial size. The rock-oyster tastes best in the pre-spawning period (November to February) but after spawning becomes shrunken and watery. The average weight of an oyster is 50 gm and large specimens can grow to 12 cm in length. Rock-oyster production in New South Wales has dropped by 30% in recent years due to the use of TBT antifouling paint on the hulls of boats moored in our estuaries and river systems. Recent legislation should help to overcome the TBT problem. See **mud-oyster**, **Pacific oyster**, **giant coxcomb-oyster**, **middens** and **antifouling paint**.

**The successful cultivation of the Sydney rock-oyster *Saccostrea cucullata* commenced in Australia around 1896.**

**symbiosis.** A relationship in which two species of dissimilar organisms live together for their mutual benefit. Symbioses have a number of common features: the association is a permanent feature of the life cycles of the organisms; the organisms are in sustained and intimate physical contact, and there is directional movement of metabolites from one organism to the other. Certain dinoflagellates, commonly called 'zooxanthellae', are common and successful algal symbionts living in the tissue of hard and soft corals, and in certain molluscs such as the giant clam, various protozoans, anemones, gorgonians, echinoderms, zoanthids and nudibranchs. See **clam** and **zooxanthellae**.

**tank.** An expression used to describe a scuba cylinder.

**tank filling compressor.** See **compressor**.

**tapestry cockle.** A bivalve mollusc and member of the family Veneridae. The tapestry cockle *Tapes dorsatus* has characteristic wavy lines on its shell. These cockles were once favoured by coastal aborigines as food, as evidenced by the large numbers of their shells found in middens along the Australian coastline. See **middens**.

**Taronga Zoo Aquarium.** The following information has been supplied by John West, Aquarium Supervisor, Taronga Zoo Aquarium:

The aquarium at Taronga Zoo in Sydney was the first to be built in the Southern Hemisphere. The aquarium was completed in two stages: The first built in 1927 incorporated three floors of exhibit tanks and a shark pool, the second stage finished two years later, in 1929, consisted of another exhibit floor which contained the tropical marine and freshwater fishes and a seal and penguin pool. As basic as this early facility was it proved to be of great importance to the knowledge of our little known Australian fishes and invertebrate life. For many years it was utilised by prominent ichthyologists from the Australian Museum as a source of newly discovered species.

A policy was developed for the display of animals at the aquarium to coincide with species management planning throughout the Zoo. The main objective was to promote conservation through education and this policy has dictated the direction that has been taken with the upgrading of the aquarium.

The aquarium displays over 220 species of endemic marine and freshwater fishes, invertebrates and reptiles within 42 exhibit tanks and incorporates a multi-thematic approach. It is hoped that this approach will encourage a better understanding of Australia's unique aquatic life and promote conservation.

As people move through the displays, dolphin, whale and fish sounds fade in and out of range. This is interspersed with running or bubbling water and the occasional crackling of invertebrate noises. One particular interactive display has a push button which reproduces some of the sounds made by the fishes on display.

The aquarium exhibits have been extensively redeveloped. The staff, using their comprehensive knowledge and experience related to the behavioural and biological needs of the many Australian fishes and invertebrates have developed naturalistic displays and more efficient life support systems.

A number of the fishes on display are rare, endangered or in one particular case, the Lake Eacham Rainbowfish is 'extinct' in the wild (only four breeding groups of the Lake Eacham Rainbowfish exist and Taronga has one). It is hoped that the display of these fishes will, through interpretive graphics and audio visual presentations, bring their environmental problems to light and encourage the public to be more aware of conservation issues.

Taronga Zoo is a biological park which is owned by the people (not a private enterprise). It is, and must continue to be a place where people come for recreation, fun and excitement. It is during their visit to the Zoo that the issues of conservation can, and must be presented through exciting new exhibits.

It is hoped that the development of an exciting new aquarium complex at Taronga Zoo will increase the public's appreciation and understanding of our unique and extensive aquatic environments and the animals that live in them. Taronga Zoo is situated at Bradleys Head Road, Mosman, NSW, the postal address is PO Box 20, Mosman, NSW 2088; phone: (02) 969 2777.

**taxonomy.** That branch of science which deals with the systematic classification of living things. See **holotype** and **scientific classification of marine organisms**.

**TBT.** An acronym for Tributyl-tin, a chemical commonly used in the manufacture of anti-fouling paints. See **anti-fouling paint**.

**tetrodotoxin.** See **pufferfish**.

**thermocline.** The interface between two layers of water of different temperatures.

**The Wheel Dive Planner.** A recreational dive planner or dive table, designed and tested specifically for recreational scuba divers. The Wheel is said to be easier and faster to use than conventional dive tables and multilevel repetitive dive profiles can be worked out easily. Available through PADI training facilities, Instructors and dive shops.

**The Wilderness Society.** See **Wilderness Society The**.

**tidal wave.** See **tsunami**.

**tide.** The periodic rise and fall of the water level in the ocean and its inlets. Tides are caused mainly by the Moon's gravitational attraction due to its closeness to Earth. The Sun and the rotation of the Earth also have an affect on tides. Tides are important when planning scuba dives as some dive sites are dangerous when the tide is changing and can only be dived safely at slack water. See **neap tide**, **spring tide** and **slack water**.

**tide-race.** A swift tidal current.

**tiger shark.** Member of the order Carcharhiniformes, family Carcharhinidae. The tiger shark *Galeocerdo cuvier* is arguably the most dangerous shark in Australian tropical waters as it is a proven people-eater. The mouth of a tiger shark contains broad serrated, cockscomb-shaped teeth in both upper and lower jaws. These sharks can grow to about six metres in length and their favourite food is turtle, including the carapace. They are sometimes called the 'garbage can of the sea' because of their habit of following ships and eating any rubbish thrown overboard. Tiger sharks are found in all tropical and subtropical oceans of the world and their common name originates from the dark tiger-like stripes on juvenile specimens; as they grow to maturity the stripes gradually fade. The tiger shark is a live bearer, sometimes producing as many as 48 darkly striped pups in a single litter. For more information on the tiger shark see Fishwatcher's Notebook in the December/January 1987 issue of *Scuba Diver* magazine. See **carapace** and **shark**.

A set of jaws from a tiger shark showing the characteristic broad serrated cockscomb-shaped teeth.

Ray Hyde feeding a tiger shark at the old Manly Marineland, Sydney.

**toadfish.** See **pufferfish**.

**Toheroa.** A New Zealand bivalve mollusc similar to the Australian pipi. See **pipi**.

**trepang.** See **sea-cucumber**.

***Tridacna maxima.*** See **clam**.

**trochus shell.** A spiral-shelled gastropod mollusc and member of the family Trochidae, also known as the button trochus. The trochus *Trochus niloticus* can be found grazing on turf algae on the seaward side of coral reefs in the Pacific and Indian oceans in depths less than 10 metres. Juveniles live in the rubble zone on the reef top immediately behind the reef edge where they find protection from predators. Trochus reach a maximum shell diameter of about 15 cm at about 12 to 15 years of age. The shell of the trochus is a luxury product which is used to produce the pearlescent sheen in nail polish and expensive car paints; in the fashion industry it is used in the manufacture of buttons and ornamental jewellery. Trochus shell buttons are in demand with quality shirtmakers as they can withstand strong detergents and frequent washing cycles much better than plastic buttons. Trochus is so prized that Indonesian fishermen have often been arrested and their boats confiscated in northern Australian waters poaching for these valuable shells.

**The trochus shell *Trochus niloticus* is commercially exploited and forms the basis of a number of small manufacturing industries.**

**tropical rock lobster.** A crustacean and member of the order Decapoda, family Palinuridae. Tropical rock lobsters are usually located around the base of reefs hiding under coral plates and bommies during the day and at night they emerge to scavenge for food such as detritus and dead fish. They are more commonly called coral (or painted) crayfish. There are about five species of tropical rock lobsters in Australian waters. The more common species include: *Panulirus versicolor, P. ornatus* and *P. longipes.* See **rock lobster** and **western rock lobster**.

**The coral crayfish *Panulirus longipes.***

The painted crayfish *Panulirus versicolor.*

The ornate crayfish *Panulirus ornatus.*

**tsunami.** Large destructive sea waves (also called tidal waves) generated by submarine upheavals such as earthquakes, volcanic eruptions and land slides. The most infamous was associated with the explosive eruption of a volcano on the island of Krakatoa in Indonesia in 1883 the resulting huge waves moved across the Indian Ocean and the Java Sea at speeds in excess of 600 km/hr, killing more than 35000 people.

***Tubipora musica.*** See **organ pipe coral**.

**tunicate.** Member of the subphylum Urochordata. The tunicates comprise three classes of animals—the Ascidiacea (ascidians or sea-squirts), the Thaliacea (salps) and the Larvacea (transparent planktonic animals). Tunicates are mostly barrel-shaped and have an external protective leathery bag or test surrounding the internal organs. The word tunicate is derived from the Latin *tunicatus,* meaning clothed with a tunic. They are solitary or colonial animals found in all seas, attached to any suitable substrate or swimming freely in the water. Most tunicates are hermaphroditic and the eggs are fertilised inside the exhalent chamber or in the open water. Being filter feeders, tunicates trap food particles in mucus produced by the pharynx. Adult tunicates are sessile but the larval forms are free-swimming and tadpole-like and feature a hollow dorsal nerve cord and notochord-like cells (which are lost on metamorphosis). Marine biologists have discovered the presence of phosphocreatine in the liver of the tunicate, *Pyura stolonifera,* suggesting the possibility of a closer relationship to vertebrates than to invertebrates. Probably the best known tunicates are the sea-squirts or cunjevoi of the marine rock platforms, so named because of their ability to squirt spouts of water into the air when stepped upon. Cunjevoi are often cut open and the meaty parts used as bait by rock fishermen. Tunicates make an interesting addition to the home marine aquarium and will reproduce in captivity if the aquarium is large enough. See **ascidian**, **compound ascidian** and **salp**.

**The tunicate *Pyura spinifera* is common to waters off southern Australia. The pink colouration is due to an encrusting sponge.**

**turtle.** Marine reptile and member of the order Chelonia. There are six species of sea turtles found in Australian waters: the green turtle, *Chelonia mydas;* the loggerhead turtle, *Caretta caretta;* the hawksbill turtle, *Eretmochelys imbricata;* the flatback turtle, *Natator depressus;* the olive ridley turtle, *Lepidochelys olivacea;* and the leatherback turtle, *Dermochelys coriacea,* which is the world's largest and most unusual turtle. The Great Barrier Reef is one of the most important sea turtle habitats in the world and the home to three of the most common species, the green, loggerhead and hawksbill turtles. Turtles at rest can remain submerged in shallow well oxygenated water for an indefinite period of time; this is due to an accessory organ of respiration—the cloaca (terminal part of the gut). Underwater the two main chambers of the cloaca are used for oxygen exchange and are efficient enough to keep the tissues of resting turtles adequately supplied. Plastic bags, which look like jellyfish to a hungry turtle, can eventually kill them by clogging up the stomach; this material is indigestible and causes slow starvation. You can help prevent unnecessary turtle deaths by collecting and disposing of any plastic debris found on the beach or seen floating in the water. A long term study of marine turtles is being carried out by various organisations using metal tags to help track migration routes. If you find a tagged turtle, dead or alive, please record the number on the tag, along with the date, time, place and circumstances and return this information to the address stamped on the tag. See **calipee**, **green turtle**, **hawksbill turtle**, **leatherback turtle** and **loggerhead turtle**.

**turtle research.** The Department of Conservation and Land Management Western Australia is undertaking a long term study of marine turtles nesting on the north and northwest coasts of the State. Adult turtles are being tagged on the trailing edge of one or both fore-flippers. Should you come across one of these tagged turtles, please record the number on the tag, along with information such as, date, time, place and circumstances and forward to: Wildlife Research Centre, PO Box 51, Wanneroo, WA 6065. See **turtle**.

**turtle weed.** A green alga and member of the division Chlorophyta. The turtle weed, *Chlorodesmis fastigiata,* is a brilliant green coloured alga often found growing on coral heads and rubble banks on reef flats and lagoon areas throughout the tropical Indo-Pacific region. Turtle weed grows in small clumps and looks like strands of green hair. It forms part of the diet of various species of turtles on the Great Barrier Reef, including the green and hawksbill varieties. See **algae**.

**The tropical green alga *Chlorodesmis fastigiata* is commonly called turtle weed.**

**twin hose regulator.** A scuba regulator that has the pressure reduction valve and the exhaust vale in a single unit that is attached to the cylinder. Twin hoses for the inlet and exhaust gases run to and from the diver's mouth. This type of regulator has the advantage of releasing the exhaust bubbles behind the diver's head, allowing clear vision at all times. By the early 1970s, most divers had replaced their twin hose regulators with single hose regulators, as most twin hose regulators could not accommodate submersible pressure gauges.

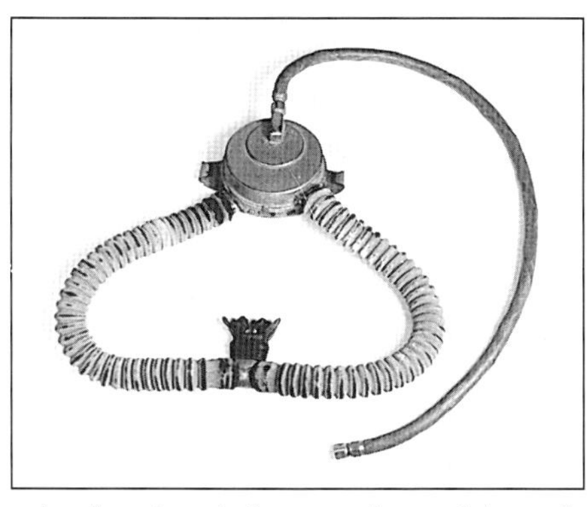

**A hand-made twin hose regulator of the early 1960s.**

**Twin hose regulators though popular in Europe, never gained a large following in Australia.**

**twin tanks.** Two scuba cylinders joined by a yoke to enable a single regulator to be attached to the twin tank combination. Twin tanks of a suitable capacity can be used by scuba divers to extend the duration of their dives and to allow an ample supply of air for decompression stops. Twin tanks are said to be more comfortable to wear and better balanced than a single large capacity cylinder. See **yoke**.

**In the early 1950s scuba equipment was scarce and expensive in the sports stores so enthusiasts made most of their own diving gear. This particular scuba outfit was made by Mr Tom Campbell of Sydney; even to the extent of moulding his own rubber regulator diaphragms. The twin cylinders are aircraft oxygen tanks which have been modified for scuba diving.**

**U-boat.** A class of German submarine active in the First and Second World Wars.

**ultramarine.** A deep blue colour.

***Ulva lactuca*.** A green alga in the division Chlorophyta, commonly called sea-lettuce. This alga with its soft, green, lettuce-like leaves, is found in great abundance in midsummer on marine rock platforms, in shallow rock pools and in estuaries. It is especially abundant in nutrient-rich areas around sewerage outflow pipes or storm water drains. It can be used for human consumption in salads and soups as long as it is collected from areas free of pollution. There are several species of *Ulva*.

**The green alga *Ulva lactuca* is commonly referred to as sea-lettuce and is quite edible.**

**undertow.** A current of water moving seaward beneath breaking surf.

**underwater.** Being or occurring under water.

**underwater camera.** See **Nikonos camera**.

**underwater camera housing.** A watertight metal or plastic container with appropriate lens port and control knobs (to facilitate focusing and aperture control) in which an ordinary camera is placed so it can be operated underwater.

**This underwater camera housing for a Nikon F2 35 mm camera was manufactured by Ikelite and is made from a tough clear plastic called Lexan®.**

**underwater communication.** Communicating underwater using various methods such as hand signals, rope signals or telephone link. Talking underwater is now a reality for the recreational diver with Aquavox, a silicon attachment for any regulator with a standard size mouthpiece. Manufactured in the USA, this device allows divers to speak in a normal voice—everyone within 10 metres will be able to hear. There are no electronic or mechanical parts and no batteries. For further information contact: AQUAVOX, INC. PO Box 612, Cape Canaveral, Florida 32920, USA.

***Underwater Geographic.*** A quarterly magazine for scuba divers and conservationists published by the famous Australian underwater photographer and author Neville Coleman. First published in 1981 as *Underwater* the name changed to *Underwater Geographic* in 1989. Contact Sea Australia Productions Pty. Ltd. PO Box 702, Springwood, Qld. 4127; phone: (07) 341 8931. See **dive magazines**.

**underwater magazines.** See **dive magazines**.

**underwater metal detector.** An electronic device that detects the presence of various metals underwater. Underwater exploration for shipwrecks and sunken treasure is gaining popularity all over the world. Treasure seekers can now purchase an Australian manufactured light weight, marine metal detector called the Hydro-Probe. Enquires should be directed to: Probe-Electric Products, PO Box 149, Berowra Heights, NSW 2082; phone: (02) 456 1404. See **magnetometer**.

**underwater photography books.** Books dedicated to the art and techniques of underwater photography. The list is not complete but the following books are helpful:
• *Beginning Underwater Photography* by Jim and Cathy Church, PO Box 80 Gilroy, California 95021, 1987.
• *Choosing and Using Underwater Strobes* by Jim and Cathy Church, PO Box 80 Gilroy, California 95021, USA, 1984.
• *How to Build your own Underwater Camera Housing,* Toggweiler; 1970.
• *Nikonos Handbook* by Jim and Cathy Church, PO Box 80 Gilroy, California 95021, USA, 1986.
• *Nikonos Photography—The Camera and System* by J. and C. Church, PO Box 80 Gilroy, California 95021, USA, 1976.
• *Mastering Underwater Photography* by Carl Roessler, William Morrow & Company, New York, 1984.
• *The Complete Guide to Underwater Modeling* by Tom Mount and Patti Schaeffer, Sea-Mount Publishing Co., 1545 N.E. 104 St., Miami Shores, FL. 33138, USA, 1984.
• *The Manual of Underwater Photography* by H. Gert De Couet and Andrew Green, Oceanimage, PO Box 41, Hackett, ACT 2602; 1989.
• *The Nikon Guide to Underwater Photography* VHS video guide from Nikon's Australian distributor Maxwell Optical Industries, PO Box 245, Pyrmont, NSW

2009; phone: (02) 660 7088.
• *Underwater Photography* by Walter Deas and Richard Rice, Ure Smith, Sydney, 1977.
• *Underwater Photography* by Charles Seaborn, Amphoto LIMP.
• *Underwater Strobe Photography* by Jim and Cathy Church, PO Box 80 Gilroy, California 95021, USA, 1976.

**underwater viewing aid.** A mechanical device, usually bucket-shaped, with a clear glass or perspex bottom which allows a surface observer a clearer view underwater. An underwater viewing aid called a SEASCOPE is being sold in Australia; this is a lightweight, hand-held underwater viewer made from plastic materials. Great for checking water clarity before getting wet. For further information contact: Stratas Tasmania Pty Ltd, 260 Macquarie St., Hobart, Tas. 7000; phone: (002) 23 1165, fax: (002) 24 0109.

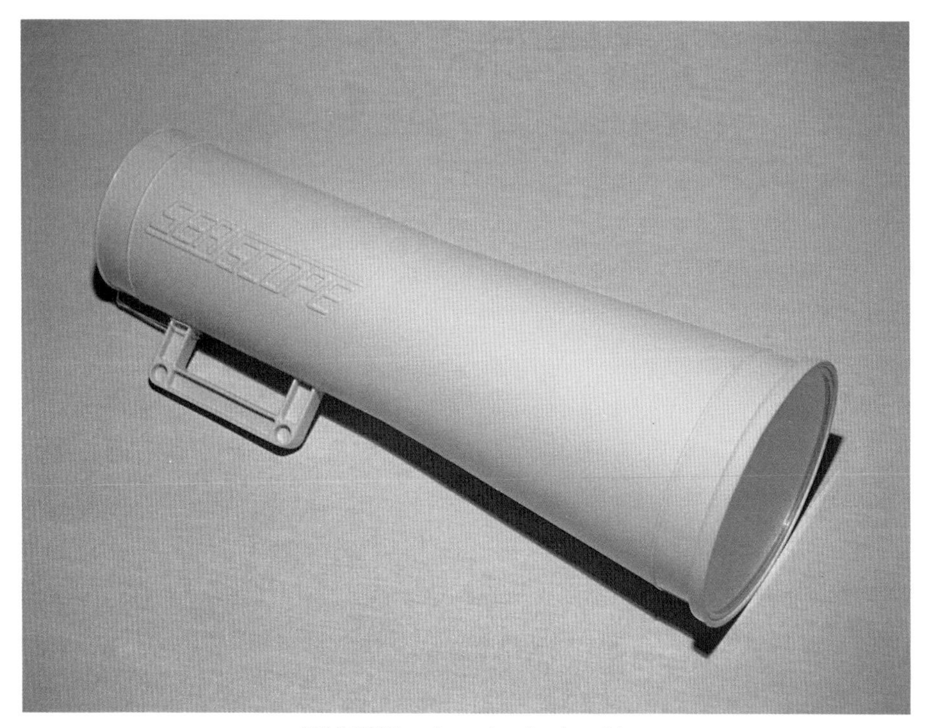


**SEASCOPE underwater viewing aid.**

**underwater watch.** See **dive watch**.

**Underwater World, Perth.** A marine aquarium 18 km north of Perth, Western Australia, opened as part of the Sorrento Quay/Hillary's Boat Harbour complex. Visitors travel on a moving walkway 98 metres in length enclosed in an underwater acrylic domed tunnel (the longest of its kind in the world),

which is situated in the main three million litre seawater tank (41 metres long x 20 metres wide). An interesting feature of the complex is Microworld, a series of smaller aquariums enclosed in booths where a visitor-operated video camera can enlarge the crabs, anemones and other sea creatures up to 12 times their normal size on a high resolution colour screen. The complex also boasts a 100 seat theatre with a continuous audio-visual presentation featuring the marine environment, and a 'touch pool' where visitors can handle live sea creatures. Underwater World, Hillarys Boat Harbour, West Coast Drive, Hillarys, WA 6025; phone: (09) 447 7500.

**upstream valve.** This simple type of demand valve, consisting of a rod and seal, which separates high pressure air from low pressure air. The seal is held closed by a spring and when the rod is tilted or opened upstream, against the air-flow, air escapes past the seal. The main advantage is that this type of valve functions well no matter what the line pressure of the regulator is. The upstream valve is also suited to surface supply equipment such as hookah units.

**upwelling.** An oceanic phenomenon wherein deepwater is drawn to the surface. Ocean currents can sweep up from great depths and these cold bottom-waters are rich in nutrients. The nutrients are taken up by the plankton which in turn provide food for huge populations of fish.

**US Navy Standard Air Decompression Tables.** A set of dive tables published by the United States Navy in 1958. The US Navy dive tables have served as the standard for recreational scuba divers for the last 30 years. In recent years a number of format rearrangements have simplified the tables for repetitive diving applications.

**Valsalva manoeuvre.** A technique used to equalise air pressure in the middle ear with that of the surrounding or ambient pressure, commonly called ear clearing. This is achieved by sealing the nostrils and blowing gently until the ears 'pop', causing pressure at the back of the throat and forcing the eustachian tubes open.

**valve.** A device that regulates the flow of air in diving equipment by starting and stopping the flow. See **upstream valve**, **downstream valve**, and **servo-assisted valve**.

**'V' distress signal sheet.** A fluorescent orange coloured sheet measuring at least 1.2 metres in height and 1.8 metres long with the letter V in the centre of the sheet. The V must measure at least 760 mm in height and 910 mm in width. The sheet should be tied with a piece of rope (lanyard) in each corner. It is used in boating emergencies to indicate you need help.

***Velella velella.*** See **by-the-wind sailor**.

**venom.** A poison secreted by certain types of animals.

**ventral.** Situated near to that side of the animal which points downwards i.e. the underside of the animal. Opposite of dorsal. See **dorsal**.

**Venus's flower basket.** A marine deepwater glass sponge, class Hexactinellida, genus Euplectella. Venus's flower basket probably has a worldwide distribution at depths of 200 to 1000 metres and is characterised by a lattice-like skeleton which looks very much like fine white porcelain. The delicate 20 to 40 cm curved tube-shaped sponge has a narrow base consisting of a tuft of straight hair-like silicaceous spicules which are buried in the substrate for support; the other end is wider and enclosed by a convex perforated sieve-plate. A pair of small shrimps are often found living inside these sponges; they enter as tiny larvae, grow into adults and because of their increased size cannot get out and so must remain imprisoned together for life. In Japan this sponge is sometimes displayed in the room where a marriage ceremony is to take place as a sign of conjugal fidelity. See **sponge**.

**vignetting.** A reduction in intensity of the light transmitted through a lens at the edges of the field of view. It may cause darkening or gradual shading at the edges of the photograph.

**violent storm.** A wind of 56-63 knots or 103-117 km/hr.

**vital capacity.** Is the maximum flow of air that can be expired by the lungs after a maximum inspiration.

**viviparous.** A term used to describe animals that bear live young. Viviparity occurs in all placental mammals.

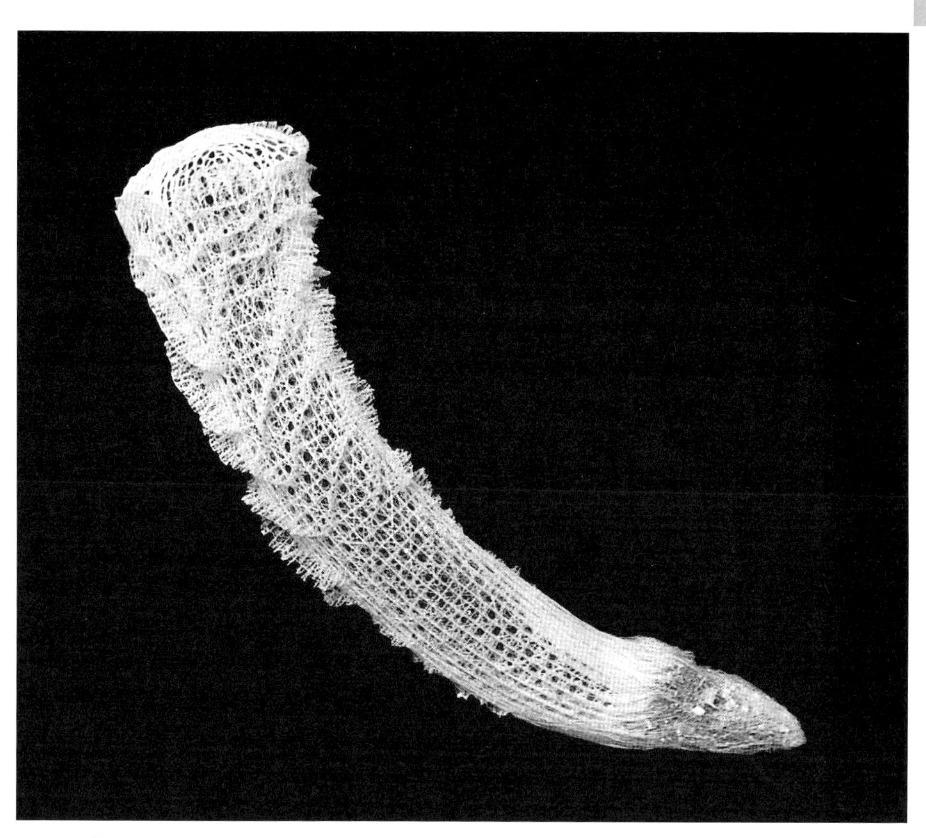

**Venus's flower basket is the only deepsea sponge well known outside scientific circles. In Europe in the 1800s it was a treasured and valuable novelty that was given a prominent position in the best room in the house.**

**Walsh, Lt Donald.** United States Navy oceanographer who, with Jacques Piccard, made the deepest human descent of 10918 metres in the bathyscaphe *Trieste* in the Marianas Trench 400 km southwest of Guam on 23 January 1960.

**watch.** See **dive watch**.

**waves.** See **sea** and **swell**.

**weight belt.** A basic item of dive equipment consisting of a number of lead weights threaded onto a belt made of artificial fibres or plastic, attached to a quick release buckle made of stainless steel or plastic. The human body has natural buoyancy in seawater; add this natural buoyancy to the already buoyant neoprene wetsuit and the need for a number of lead weights to achieve neutral buoyancy becomes obvious. If a diver releases a weight belt for any reason it should be held well clear of the body before it is let go to make sure it does not snag equipment when it is released.

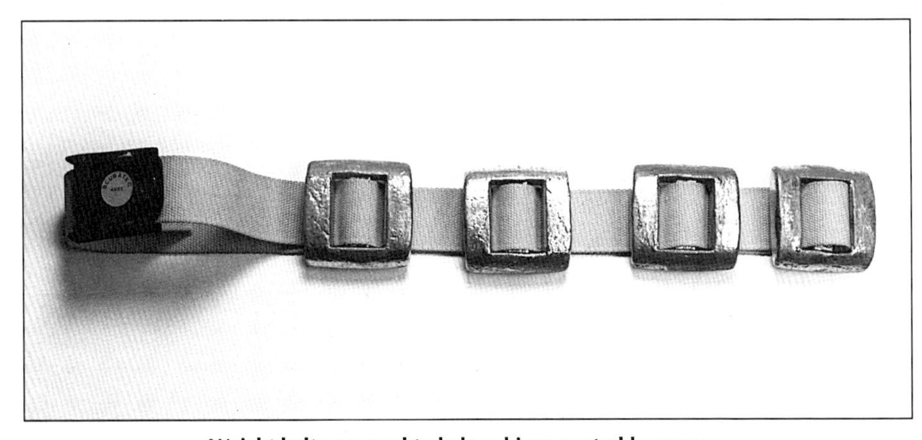

**Weight belts are used to help achieve neutral buoyancy.**

**western rock lobster.** A crustacean and member of the order Decapoda, family Palinuridae. Western rock lobster is the official name but they are still commonly called 'crays' or 'crayfish'. Adult specimens of the western rock lobster *Panulirus cygnus* may reach 4.5 kg in weight, but most are caught in the 0.5–1.0 kg weight range. The body colour varies from creamy white to brick red and purple-black. The western rock lobster also has a characteristic white spot at the base of each spine on the tail and is common under ledges, in caves and amongst coral on the limestone reefs off the Western Australia coast between North West Cape in the north and Cape Leeuwin in the south. Found at depths of up to 200 metres it is, however, most common in water shallower than 60 metres. The fishing season extends from the 15 November to 30 June each year and the minimum legal size is 76 mm carapace or head length. There are restrictions on taking rock lobster by diving but as the regulations vary from State to State they are outside the scope of this publication. The western rock

lobster industry is the most valuable single species commercial fishery in Australia. A total of nine species of rock lobster are found in Australian waters. See **rock lobster** and **tropical rock lobster**.

The western rock lobster *Panulirus cygnus.*

**wetsuit.** An underwater exposure suit made of neoprene material that hugs the body contours and allows a thin layer of water to be held between the suit and the skin. This layer of water is quickly heated by body temperature and allows the diver to remain for a longer period in cold water than a diver without a wetsuit. Wetsuits also provide protection from marine hazards such as sharp coral, barnacles, hydroids, jellyfish and a host of other stinging creatures. See **lycra suit** and **dry suit**.

Modern wetsuits are available in a multitude of designer colours.

**whale.** A marine mammal and member of the order Cetacea. Whales are classified into two main categories, suborder Odontoceti and suborder Mysticeti depending on whether they have baleen plates or teeth. The first and smaller group comprises the toothed whales (suborder Odontoceti) and includes the sperm whale, beaked whale, narwhal, beluga, killer whale, pilot whale, dolphins and porpoises. Being carnivorous and active predators these cetaceans include a variety of food in their diet including: cuttlefish, squid, fishes, and in some cases seals and other whales. Sperm whales live in the warmer oceans and are rarely seen inshore, preferring the deeper ocean waters where they dive to at least 1100 metres to seek their prey, giant squid and cuttlefish. The second group, the baleen or whalebone whales (suborder Mysticeti) includes: the blue, fin, sei, minke, humpback, bowhead, right and grey whales. These whales use fringed horny baleen plates attached to their upper jaw to strain plankton from the water. Baleen whales feed in the Arctic and Antarctic oceans during summer where krill is abundant, and migrate to the warmer oceans in the winter where cows give birth and the adult whales mate. The largest animal that has ever lived on this planet is the blue whale. It can grow to 33 metres in length and 160 tonnes in weight on a diet of up to three tonnes of krill per day. The tongue of a blue whale can weigh as much as an elephant. Worldwide there are about 80 species of whales with more than 50 species being found in Great Barrier Reef waters. See **dolphin** and **humpback whale**.

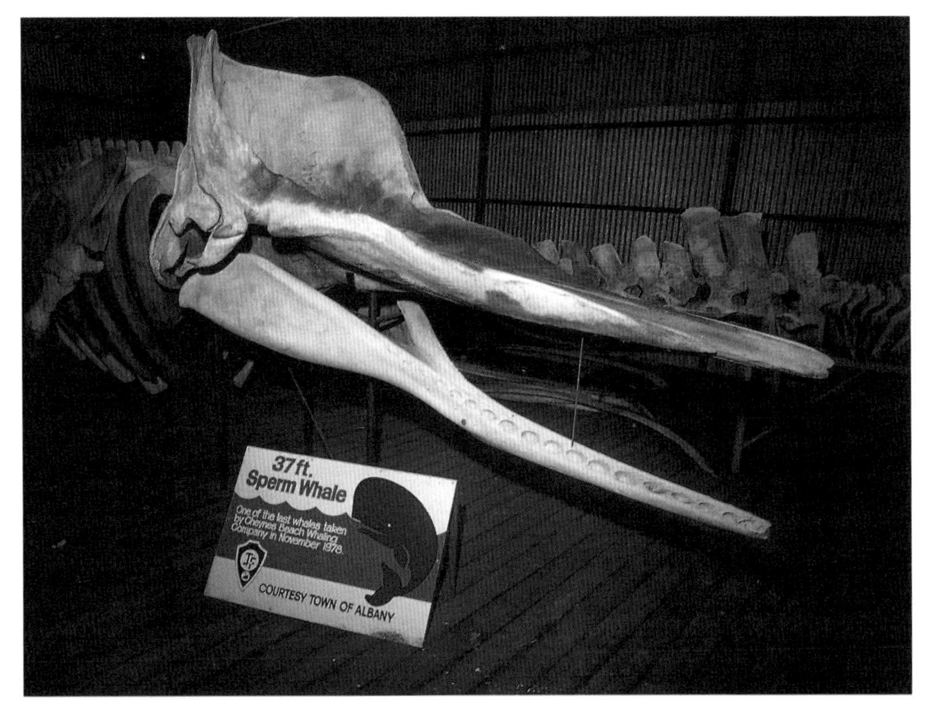


**Prior to the closure of the whaling station in Albany, Western Australia in 1978, whale bones were cooked by steaming and ground into a meal which was sold as stock feed.**

**whale products.** Commercial products made from the body of whales. Before oil from petroleum was freely available whales were mainly harvested for their blubber which was boiled down to yield high quality lubricants and transmission oils. Other uses of whale oil were in the plastics and chemical industries, steel hardening, leather dressing, rust preventatives, lamp oil, and as an additive in the manufacture of gelatin, cosmetics, wool and textiles. Whale bones as well as teeth were used as ornaments and for scrimshaw. Spermaceti wax found in the head of sperm whales was used for candles, hand creams, ointments, polishes and soaps. Whale meal was also used for stock feed and fertiliser. Ambergris from the stomach of sperm whales was used as a fixative in expensive perfumes and soaps. See **ambergris, scrimshaw** and **spermaceti**.

**These storage tanks at the old Albany whaling station, Western Australia were built to hold whale oil. A single sperm whale of 13 m in length and 30 tonnes in weight yields an average of seven tonnes of oil.**

**whaler shark.** Member of the order Carcharhiniformes, family Carcharhinidae. The genus *Carcharhinus* contains the whaler sharks which have been responsible for many attacks on humans. The common species found in Australian waters include: *Carcharhinus brachyurus* the bronze whaler; *C. obscurus* the black whaler; *C. melanopterus* the blacktip reef whaler; and *C. amblyrhynchos* the grey reef whaler. A few species of whaler sharks occur in rivers as far as

3700 km from the sea and have been responsible for deaths in Australia, the United States of America, India, Africa and the Amazon. See **shark**.

**This grey reef whaler *Carcharhinus amblyrhynchos* was killed with a .303 calibre powerhead fired from a speargun.**

**whale shark.** Member of the order Orectolobiformes, family Rhiniodontidae. The whale shark *Rhiniodon typus* is the largest fish in the world and grows to 13 metres in length and is completely harmless to humans. It is a sluggish swimmer and filters plankton, small fishes, squid and other plants and animals from the water with its gill rakers. It is oviparous and lays an egg case about 30 cm long and 12 cm wide. The newborn sharks are about 35 cm long. Whale sharks are found in all tropical and warm temperate oceans and people have on occasions walked on their backs. Scuba divers have sometimes been able to hitch a ride by holding onto the large dorsal fin or the top edge of the upper jaw. See **oviparous** and **shark**.

**whistle.** A noise-producing metal or plastic device held in the mouth and blown to attract attention. A whistle attached to your buoyancy vest is an effective way of attracting attention when in the water as it can be heard over 600 metres away if an observer or rescuer turns off the boat's engine. Whistles were commonly attached to horse collar buoyancy control devices (BCs) but are rarely seen today on modern BCs. See **Safety Sausage**.

**white pointer shark.** Member of the order Lamniformes, family Lamnidae. The white pointer, *Carcharodon carcharias,* is also known as the great white

and the white death. These sharks are found in all tropical to temperate seas and have the unusual ability of lifting their heads from the water. This behaviour enables them to search for prey—seals and sea-lions which live on rocky ledges —and hopefully scare them into the water. This shark has been responsible for more documented human fatalities than any other species. The largest specimens may reach eight metres in length. See **shark**.

**This white pointer shark *Carcharodon carcharias* has been preserved in a large tank of alcohol for exhibition and scientific study.**

**whitetip reef shark.** Member of the order Carcharhiniformes, family Carcharhinidae. The whitetip reef shark, *Triaenodon obesus,* is one of the most common sharks on the coral reefs of the tropical Indo-Pacific region and can grow to 2.3 metres in length, feeding mainly at night on small fishes, eels and octopus. During the day whitetips spend most of the time resting in small caves. This species bears live young. Due to their small size they pose little danger to scuba divers. See **shark**.

**The whitetip reef shark has a characteristic white tip on the first dorsal fin and the tail.**

**white water.** Water in which air bubbles are suspended or where the surface is disturbed by wave action.

**Wilderness Society, The.** A Society dedicated to saving Australia's wilderness areas. The following information has been supplied by The Wilderness Society:

Following the drowning of Lake Pedder National Park in South-West Tasmania by a hydro-electric scheme, a small group of people met in 1976 to form a society dedicated to the protection of Tasmania's wilderness. In 1978, the Society began a campaign in earnest to stop the Hydro-electric Commission's plans to flood the Franklin River. That campaign was waged over five years and concluded with the momentous High Court decision in July 1983 which prevented the construction of the dam.

After 1983 The Tasmanian Wilderness Society altered its name and structure to become The Wilderness Society. In just over 10 years, millions of hectares of Australia's wilderness have been saved through the efforts of The

Wilderness Society. From the Franklin River and Kakadu to the Daintree rainforests and New South Wales wilderness areas, the Wilderness Society has been at the forefront of campaigns to protect Australia's wilderness.

Over recent years the Society has produced award winning films, outstanding photographic books, organised hundreds of public meetings and displays and spoken to thousands of groups. For schools an extensive program of Wilderness education kits is being produced.

Around the country Wilderness Society branches are working vigorously on wilderness protection but much remains to be done. The Wilderness Society has a network of 15 branches and shops around the country. Enquiries: The Wilderness Society, 130 Davey Street, Hobart, Tas. 7000; National toll-free number phone: 008-030 641.

**Wild Life Preservation Society of Australia.** A society dedicated to the protection of plant and animal life. The following information has been supplied by the Wild Life Preservation Society of Australia:

Started in 1909 this Society is the oldest purely conservation group in Australia and possibly one of the oldest in the world. Most of the time it has worked as a lobbying organisation, its early work being devoted to saving the koala and egrets as well as stimulating government concern about laws to protect many aspects of the environment, particularly plants and animals.

After the First World War the Society began monthly lectures in Sydney but although these flourished for a time they have now been reduced to lectures every three months. Often these are given by world famous figures. The president is available to travel anywhere in Australia to help conservation causes, the only demand being that the organisations concerned meet travelling and accommodation costs. No fees are charged for this service and over the last ten years all States have been visited and some trips made overseas. Contact address: Wild Life Preservation Society of Australia, Box 3428, GPO Sydney, NSW 2001; phone: (043) 434 708.

**Wildlife Preservation Society of Queensland (WPSQ).** The following information was supplied by the Wildlife Preservation Society of Queensland:

The Wildlife Preservation Society of Queensland is one of Queensland's oldest and most respected environmental groups. Founded in Brisbane in 1962, the Society has been involved over the years in some of Queensland's most important conservation issues. It campaigned during the 1960s for the protection of the Great Barrier Reef and for the establishment of the Cooloola National Park. In the 1970s it worked to have national parks declared in the State's northern rainforests and successfully opposed sandmining on Fraser, Moreton and Stradbroke Islands. In the 1980s it halted development plans for Lindeman Island and mining at Shelburne Bay and lobbied to have the northern rainforests World Heritage listed. It was also involved in the attemp to save vital habitat caves of Mt Etna's ghost bats.

The Society is a non-profit community-based organisation with over 25 branches throught the State. Our aims are the conservation of the flora and fauna of Australia and their habitats.

As a major national environmental education initiative, the Society produces Australia's premier wildlife magazine *Wildlife Australia.* Profits from the sale of the magazine are ploughed back into environmental conservation work. Contact address: Level 4, 160 Edward Street, Brisbane Qld 4000; phone (07) 221 0194, fax: (07) 221 0701.

**wind.** Air in natural motion. A wind is known by the direction from which it blows, in direct contrast to water movements (currents, tidal streams and wave sets) which are named for the direction to which they are moving. Changes in the direction of the wind are most noticeable after the passage of a pressure system or front. If the observer is facing into the wind and it changes direction to the observer's left the wind is said to be 'backing'. Conversely, if the wind changes to the observer's right, it is said to be 'veering'.

A rule of thumb weather guide:

- If the wind follows the sun, from the east in the morning to the west in the evening, this usually means good weather.
- If the wind increases in the morning and decreases in the afternoon, this can mean good weather.
- If the wind suddenly changes direction after several days of blowing in the same direction, this can mean bad weather.
- If the wind increases in the afternoon or evening, this can mean rain or a storm are likely.

**windward.** Towards the wind, the side from which the wind blows, opposed to leeward.

**wobbegong shark.** Member of the order Orectolobiformes, family Orectolobidae. There are six species of wobbegongs and they are all slow

**The spotted wobbegong *Orectolobus maculatus* is well camouflaged and it is obvious why it is commonly called a carpet shark.**

moving, bottom dwelling sharks also known as carpet sharks. Various species are found in every State of Australia and throughout the Indo-Pacific region. Some of the larger species such as *Orectolobus maculatus* commonly called the spotted wobbegong may reach three metres in length, and have been known to take a bite at careless scuba divers who tease them by pulling their tail, or who carry speared fish on their person. Wobbegong sharks are the sharks most commonly encountered by scuba divers in Australian temperate waters and bite more divers than any other species. For more information see Fishwatcher's Notebook in the April/May 1989 issue of *Scuba Diver* magazine.

**working pressure.** The maximum pressure to which a scuba cylinder should be filled for use. This pressure rating is stamped on the neck of the cylinder by the manufacturer. The working pressure varies between brands, sizes and whether constructed from aluminium or special steels.

**World Underwater Federation.** Synonymous with the Confederation Mondiale Des Activites Subaquatiques. See **Confederation Mondiale Des Activites Subaquatiques**.

**wreck diving.** Scuba diving on sunken vessels. With planning, a wreck dive can be enjoyed by most scuba divers. If you plan to investigate the interior of a large wreck it is recommended that a specialist wreck diving course be undertaken for safety. These courses are offered by most dive shops. There are many shipwrecks around the Australian coastline representing various phases of maritime history and technology. Wreck divers should ensure their dive knife is sharp as there can be many entrapments on wrecks, including snagged fishing nets, lines and ropes. An underwater torch is essential as is a line and reel for safe entry and exit from inside the wreck. Artifacts should be left where they are found. It is illegal to remove artifacts from any pre-1900 wreck in Australian waters or wrecks which have been declared historic under the Historic Shipwrecks Act of 1976.

Diving the shipwreck *Coral Queen* off Madang, Papua New Guinea.

**Xenia.** A genus of soft coral and member of the order Alcyonacea. *Xenia* is grey-brown in colour, the colonies are small and sparsely branched with non-retractile polyps that rely on the photosynthetic productivity of their symbiotic zooxanthellae for most of their nutrients. *Xenia* lacks the rigid limestone skeleton of the hard reef building corals. See **soft coral**.

**Xiphosura.** Order of the phylum Chelicerata. The order Xiphosura contains the horseshoe crabs—also called king crabs (family Limulidae) which are bottom dwellers and are the sole surviving members of a group of ancient marine creatures which were common in the oceans in the Ordovician period about four to five hundred million years ago. The front and middle parts of the body form a spiked horseshoe-shaped jointed shield and attached to the rear section of the body is a long tail or spine. Horseshoe crabs live on soft sand or muddy substrates which they burrow through to find their omnivorous diet, consisting mainly of molluscs, worms and algae which they swallow whole. Horseshoe crabs can swim by flipping over on their backs and paddling with their legs and right themselves using their long tail spike. See **Chelicerata**.

**Horseshoe crabs have heavily armoured bodies which may reach 60 cm in length.**

**yabbie.** See **yabby.**

**yabby.**[1] Several species of marine crustaceans of the family Thalassinidea. The yabby *Callianassa australiensis* is a common species in estuaries in New South Wales and Queensland. Yabbies are white or pink in colour, prawn-like in appearance and live in burrows over estuarine sand/mud flats where they are collected by fishermen for bait using yabby pumps. An older but slower method of capture works quite well on eel-grass mud flats at low tide. This technique requires a hole to be made in the mud with a puddling action of the feet. The yabbies are forced out of their burrows by the puddling action and can then be scooped out of the water-filled hole. Yabbies are also called nippers. See **Zostera sp**.

A male and female yabby *Callianassa australiensis*. The male has a large left nipper (claw).

**yabby.**[2] An Australian freshwater crayfish of the genus *Cherax*.

**yabby pump.** A device for catching yabbies, consisting of a stainless steel tube about 60 cm long and 50 mm in diameter containing a plunger attached to a handle. The open end of the tube is thrust into the sand and the plunger is drawn upwards. When removed from the sand the contents of the tube is expelled and any yabbies present are collected and used for bait. See **yabby**.[1]

**yellowfin tuna.** A large pelagic species of fish found worldwide in temperate waters. The yellowfin tuna *Thunnus albacares* is of economic and sporting importance, thousands of tons are taken each year. Portion of the catch is used in the preparation of sashimi. See **sashimi**.

The yellowfin tuna *Thunnus albacares* may reach 100 kg in weight.

**yellowtail scad.** A fish and member of the family Carangidae. *Trachurus novaezelandiae,* commonly called yellowtail or yellowtail scad, is very common in estuarine and coastal waters of Australia's southern half. Yellowtail form large schools and can reach 33 cm in length. These fish make very good bait and are commonly caught from wharves. Scuba divers often encounter large schools on coastal reefs.

Juvenile yellowtail scad *Trachurus novaezelandiae* are often found under the large domed bell of the jellyfish *Catostylus mosacicus.*

**yoke.** A metal device for joining two scuba cylinders together so they can be used with one regulator.

**yolk.** A store of food material, such as fat and protein, present inside the eggs of the majority of animals. See **yolk-sac**.

**yolk-sac.** A sac containing yolk which hangs from the ventral surface of vertebrate embryos in the class Chondrichthys—the cartilaginous fishes (sharks, rays and skates). See **ventral** and **yolk**.

**zebra lionfish.** Member of the family Scorpaenidae. The zebra lionfish *Dendrochirus zebra* has a number of common names, including butterfly-cod, because of slow graceful undulations of the pectoral fins, and firefish, because of the painful sting which can be caused by touching the long venomous spines on the dorsal, anal and pelvic fins. Zebra lionfish make good aquarium specimens but care should be exercised

A juvenile zebra lionfish *Dendrochirus zebra.*

as they have very large mouths in proportion to their size and will eat many of the other fishes in the aquarium. For more information see Fishwatcher's Notebook in the August/September 1987 issue of *Scuba Diver* magazine.

**zebra shark.** Member of the order Orectolobiformes, family Stegostomatidae. The zebra shark *Stegostoma fasciatum* is also known as the leopard shark, and is a harmless bottom-dwelling species. Juvenile zebra sharks are black and carry a series of yellow or white stripes; as they mature these stripes give way to a profusion of brown spots on a yellowish brown background. Growing to three metres in length, they are found on shallow sandy reef flats in northern Western Australia, the Northern Territory, Queensland and northern New South Wales. Feeding mainly at night, the zebra shark's diet consists of crustaceans, molluscs and small fishes. See **shark**.

**The zebra or leopard shark *Stegostoma fasciatum* is a harmless bottom-dwelling species.**

**zinc sulphate.** A chemical compound ($ZNSO_4.7H_2O$) used in marine aquariums to combat skin parasites, such as *Oodinium, Helminthiasis,* and *Trichodina* which commonly infect fishes. This compound is also effective against fungal infections on fish.

**Zoantharia.** Subclass of the Anthozoa which includes four orders: Zoanthidea (the zoanthids); Corallimorpharia (the solitary and colonial anemones); Actiniaria (the sea-anemones) and Scleractinia (the hard corals). See **zoanthid**, **jewel anemone**, **sea-anemone** and **hard coral**.

**zoanthid.** A small anemone-like coelenterate and member of the order Zoanthiniaria (= Zoanthidae). Superficially zoanthids resemble small anemones but most are colonial, united at their bases, and live as encrusting organisms. Zoanthids also have a marginal circlet of unbranched tentacles and live attached to rocks, sponges, hydroids, gorgonians, corals, bryozoans, tube worms and submerged objects. In New South Wales coastal waters zoanthids can be found growing as encrusting mats lining the underwater surfaces of rock pools and some of the colonies may be exposed to the direct rays of the sun at low tide but survive well in this hostile environment.

**This zoanthid of the genus *Palythoa* was photographed at Rottnest Island, Western Australia.**

**zoogeography.** The study of the geographical distribution of animals.

**zooid.** An individual member of a colony of animals which are joined together, e.g. a polyp.

**zoology.** The scientific study of animals.

**zoophyte.** An old term once used for various animals that are like plant-like in appearance e.g. corals and sea-anemones.

**zooplankton.** Animal plankton which float, swim or drift in the sea from the surface to the deepest depths. Zooplankton includes the single-celled Protozoa, and the larval stages of the Mollusca, Echinodermata, Crustacea, Urochordata, Coelenterata and the various larval worms and fishes. See **plankton**.

**zootoxin.** Any poison excreted by an animal.

**zooxanthellae.** Unicellular, endosymbiotic algae (dinoflagellates) living within cells in the tissues of corals, clams, anemones, zoanthids, gorgonians, nudibranchs, etc. Zooxanthellae live in a symbiotic relationship with the host invertebrate as both partners benefit from each other. Photosynthetic products from zooxanthellae supply nutrients and help the host organism such as corals to grow in nutrient deficient tropical waters, while the zooxanthellae use waste products, such as carbon dioxide, excreted by the coral, for their own metabolic needs.

***Zostera* sp.** Marine flowering plant (Angiosperm) and member of the genus *Zostera.* There are a number of species in this genus including *Zostera muelleri* and *Zostera capricorni.* They are commonly called eel-grass because of their narrow grass-like leaves which rise from short stems that in turn are attached to a creeping rhizome under the sand. Eel-grasses grow to form expansive mats on sand or mud banks in estuarine and shallow ocean environments which are important nursery areas for fishes and invertebrates. Many species of fish, molluscs and echinoderms find food and shelter among *Zostera* leaves. See **dugong-grass** and **sea-grass**.



***PHOTOGRAPHIC CREDITS***

*Special thanks to Paul De Rome for the photograph of the marine stinging sponge* Neofibularia mordens *and to Dr Martin Riddle for the isopod and goose barnacle photographs. All other photographs by Robert Berthold.*

***PHOTOGRAPHIC EQUIPMENT***

- *Nikon F2 with DA2 action finder and 55 mm micro lens.*
- *Ikelite underwater camera housing.*
- *Nikonos II & V underwater cameras with 35 mm & 15 mm lenses.*
- *Aqua sea 140 strobe.*
- *SB-103 Nikonos strobe.*
- *Film used was mainly Kodachrome 64 and T-MAX 100.*